CHILDREN OF
PROMETHEUS

CHILDREN OF PROMETHEUS

*The Accelerating Pace
of Human Evolution*

CHRISTOPHER WILLS

HELIX BOOKS

PERSEUS BOOKS
Reading, Massachusetts

Many of the designations used by manufacturers and sellers to distinguish their products are claimed as trademarks. Where those designations appear in this book and Perseus Books was aware of a trademark claim, the designations have been printed in initial capital letters.

Library of Congress Catalog Card Number: 98–86420

ISBN 0-7382-0003-4

Jacket design by Suzanne Heiser
Text design by Karen Savary
Set in 11/14 Janson Text by CIP

1 2 3 4 5 6 7 8 9-DOH-0201009998

First printing, September 1998

Find Perseus Books on the World Wide Web at
http://www.aw.com/gb/

*To my daughter Anne Marie, who will see more
of the future than I will, and who will
help her patients to survive it.*

. . . [W]hat other hand than mine
Gave these young Gods in fulness all their gifts?
. . . [L]ike forms
Of phantom-dreams, throughout their life's whole length
They muddled all at random, did not know
Houses of brick that catch the sunlight's warmth,
Nor yet the works of carpentry. They dwelt
In hollowed holes, like swarms of tiny ants,
In sunless depths of caverns; and they had
No certain signs of winter, nor of spring
Flower-laden, nor of summer with her fruits . . .
Until I showed the risings of the stars,
And settings hard to recognize. And I
Found Number for them, chief devise of all,
Groupings of letters, Memory's handmaid that,
And mother of the Muses. And I first
Bound in the yoke wild steeds, submissive made
To the collar or men's limbs, that so
They might in man's place bear his greatest toils. . . .

AESCHYLUS, *Prometheus Bound* (460 B.C.)

TR. E. H. PLUMPTRE, 1868

CONTENTS

❖

INTRODUCTION

◈

*So dominant is the forest that it is said to be possible for
an orang-outang to travel from the south to the north
of Borneo without descending from the tree-tops. His
only barrier would be the big rivers and, since the
majority of these run north and south, they would
merely prevent his spread longitudinally east and west.*

PATRICK M. SYNGE, *Beauty in Borneo* (1932)

Myth ascribes to Prometheus the gift of fire to humans. His presumptuous transfer of this power from the gods to mere mortals caused him to suffer a terrible fate. The bards first sang that tale at least four thousand years ago: they were trying to provide an explanation, in terms their listeners could understand, for why human beings have taken a path that is so different from those taken by all other animals.

It subtracts nothing from the power of the Promethean legend to realize that our own path has been so different, not because of fire stolen from heaven, but because of the forces of evolution. After all, it was those same evolutionary forces that gave us the ability to devise the Promethean legend itself, along with all the other tales, poems, theories, and inventions that have transformed our existence in the subsequent four thousand years.

Evolution has provided us with the power to accomplish all that change. But are we still evolving, and if so how? What might we be evolving into? These questions have fascinated biologists ever since Charles Darwin. In this book I will explore some surprising recent answers and show not only that we are still evolving, but that our

1

evolution—particularly the evolution of our minds—is actually proceeding at an accelerating pace.

In this respect humans are dramatically different from our close primate relatives, such as chimpanzees, gorillas, and orangutans. They are, without exception, evolving at a much slower rate. The ancient Greeks would have said that they never received the Promethean spark. We now say that, whatever combination of chance and unusual environmental circumstances caused us to start us down our path, that combination never arose in the environment of our close relatives. The scientific revolution has allowed us to explain that process in terms that are a little more specific, and a great deal less poetic, than the Promethean legend.

Whatever that evolutionary spark might have been, why did our primate relatives not receive it? To explore this question, let me begin by examining a close animal relative of ours, one that has undergone very little change over the past two million years—but that is changing now, primarily because of the impact that our own species has had on its remote and isolated world. This relative of ours is evolving slowly, in the way that Darwin first envisioned. Indeed, such slow evolution is the norm among the animals and plants of our planet—humans and a few others are dramatic exceptions. So to make the contrast vivid, let us begin with an example of the norm.

Borneo is the largest and one of the most striking of the seventeen thousand islands that make up the vast Indonesian archipelago. The north and center of the island are dominated by high mountains, but as I flew toward it recently across the Java Sea from the southwest, the mountains were quite invisible, lying far below the misty horizon. My first glimpse of the island was a flat green plain, cut here and there by the threads of rivers and streams.

Travel through this part of Borneo is still primarily by water. A few settlements cling to the edges of the rivers, and large hand-built Dayak boats move majestically between them. In the riverside town of Pangkalanbun, we hired a little boat with a cheerful Dayak crew to take us further inland to nearby Tanjung Puting National Park.

The park occupies a blunt peninsula jutting out from the coast. Here the ground is low and swampy, and the rivers overflow their banks during the rainy season, flooding the forest for hundreds of feet on either side. The area is ideal for tree-dwelling primates who rarely need to come down to ground level. Bands of long-tailed macaques swing through the trees, and even from a boat it is easy to catch glimpses of the tangled branches of old orangutan nests.

At one point a troupe of proboscis monkeys leaped from a tree that hung out over the river and swam furiously across. Their webbed fingers and toes make them some of the most impressive swimmers of the primate world; the danger from crocodiles lends an extra urgency to their natatorial escapades. As they threshed through the water, it appeared to me that they were using a stroke remarkably like the modern crawl—if so, an excellent example of evolutionary convergence with a human behavior!

We headed away from the muddied main river, which has been badly polluted by mercury and tailings from gold mining farther upstream, and quickly found ourselves in untouched wilderness. Now the stream flowed clear, its waters stained a deep orange color from dissolved vegetable matter. It was here, in 1971, that Canadian anthropologist Birute Galdikas established her first camp. The site has since grown into a substantial scientific settlement, built on fairly high ground in the heart of the forest, several hundred meters from the river. For years Galdikas and her coworkers were forced to wade through the leech-infested, waist-high mud and water to get to their settlement, but the camp is now connected to the river by a long wooden causeway.

The camp is named after Louis Leakey, the brilliant paleontologist who founded the study of fossil humans in East Africa. Leakey realized that primate behavior might give important clues to the behavior of our ancestors. He encouraged Galdikas, along with Jane Goodall and the late Dian Fossey, to study the behavior of our closest living relatives in the wild. Goodall chose chimpanzees, and Fossey chose gorillas. Galdikas, the third of Leakey's recruits, turned to the least known of this group of species, the orangutans.

We followed her into the forest in order to observe the orangs as they swung through the high treetops, as much as sixty meters above the ground, and showered us with fragments of unripe durian fruit. Her meticulous field observations have shown that these primates can feed on four hundred different kinds of fruit, so that they are always able to find food somewhere in a healthy and diverse forest. Their survival, however, depends absolutely on the maintenance of that diversity, for in the forests of Southeast Asia many tree species flower and fruit suddenly and unpredictably. The massive outpouring of fruit and seeds overwhelms the birds and animals that feed on them, so that some seeds escape. Orangutans range widely through the forest, searching for these sudden and unpredictable bonanzas.

At least in Tanjung Puting the orangutans can still pursue this ancient way of life. Primarily through Galdikas's efforts, the entire peninsula has now been saved from massive logging—although poachers still nibble at the edges of the park. She has been repeatedly threatened as a result of her efforts and was once briefly kidnapped. It is becoming grimly obvious that hers is ultimately a losing battle, but it is one that is nonetheless worth fighting.

AN EVOLUTIONARY FEEDBACK LOOP

Although Borneo is still one of the most remote and exotic places in the world, it is starting to feel the impact of the modern world. By visiting the island before it changed utterly, I hoped to fit some further pieces into the great mosaic of impressions and ideas about human origins that evolutionary biologists have been putting together ever since Darwin. My ultimate goal was an immodest one—I wanted nothing less than satisfactory answers to some of the great remaining puzzles that confront students of human evolution.

The first of these puzzles has to do with the differences between human beings and our primate relatives, particularly the great apes. Why are they so huge? Why did orangutans, for example, stay unchanged in their forest during much of the time when the human lineage was evolving into the most remarkable set of creatures the planet has ever seen? After all, we too started our evolutionary history in a forest, although it was in Africa rather than Asia. Why were our own ancestors catapulted in new evolutionary directions when the ancestors of the orangutans were not?

A few years earlier I had proposed a partial answer to this question in a book called *The Runaway Brain*. I had suggested that during the last several million years our ancestors were caught up in a runaway evolutionary process. When genetic changes led to changes in their brains and bodies, some of them were selected for, as our ancestors' own activities created an ever more complex and ever more human-oriented environment.

All these genetic and environmental changes have tended to reinforce each other in a feedback loop. Any cultural or environmental change would have selected for individuals who happened to carry genes that made them best able to take advantage of it. As a result, some of them (or their progeny) would have been clever enough to produce still more cultural and environmental changes. This would lead to even more selection, not only for new genetic variants that had arisen by mutation but also for new combinations of old genes that are produced by the genetic shuffling known as recombination that takes place each generation.

Once humans were caught up by this feedback loop, it would have been difficult for them to stop or reverse it. But the nature of the trigger that started our ancestors on this runaway course remains elusive. At the moment, among students of human origins the favored candidate for such a trigger is a sudden alteration in the climate. The problem with this, as was recently pointed out by Mark Ridley in a book review in *The New York Times*, is that "there is (as the detective Poirot would say) too much evidence: too much climatic change . . . some climatic change or other will almost always be associated with any evolutionary event." Sudden climatic alterations are happening all the time, which means that other primates should have been just as affected by them as our own ancestors were.

If it was indeed a climatic change that served as a trigger, then our own ancestors reacted to it in a different way from their great ape relatives. When the ancestors of apes were confronted by a change—perhaps something that caused the forests in which they lived to shrink in size—they did not adapt to that new condition. Instead, they were simply forced to reduce their own range. Rather than conquering their environment, they merely reacted to it.

The initial trigger that started our human ancestors on their unique course must have had about it at least a hint of new worlds to conquer. As many people have suggested, our forebears may have had to invade some new environment because the old one had essentially disappeared or because they had been driven from it. If the conquest of that new environment happened to give some clear advantage to upright posture, tool use, communication skills, or all three, then this shift would have set them on their way. From that point on they never looked back. The choice might initially have been a matter of life or-death, but it eventually turned into a wonderful evolutionary opportunity.

Soon after I arrived in Borneo, I realized why this process had not happened to the orangs: They cannot escape the forest. No other ecosystem in the area can provide them with the abundant plant food that they need. The Borneo orangs are still where they always were simply because they have never needed to cope with an ecological upheaval that forced them to live elsewhere. Had their forest vanished, they too might have joined us on that evolutionary escalator.

The second evolutionary question about which I hoped to find clues in Borneo is a more immediate one, though it is related in many ways to the first: Are we, unlike our primate relatives, still caught up in that feedback loop? Are we still evolving, and if we are, into what?

To answer this question, I have drawn from much research into both human evolution and the evolution of the pathogenic organisms that prey on us, along with many discussions with scientists who are examining behavioral and social change, how our genes work, and the interactions between us and the many other species with which we share the planet. The picture that emerges is one in which our physical evolution is certainly continuing, perhaps at roughly the same rate it always has, although certainly too slowly for us to perceive any change in our lifetimes. But at the same time, exploding cultural change is laying the groundwork for a far more rapid evolution of our mental capabilities in the future. This evolution, too, will not be visible in a single human lifetime, although it has proceeded at an accelerating pace during our evolutionary history and is likely to take place even more quickly in the centuries ahead.

Rapid environmental change causes the evolution of the organisms that are caught up in it to accelerate. By this principle, there can be no doubt that we are still evolving: we are altering our environment, and invading new ones, at an accelerating speed. Sometimes the alterations have all the earmarks of disaster. In the year since I left Borneo, for example, dramatic changes have taken place on the island. During the summer and fall of 1997, dozens of massive forest fires raced through logging-damaged areas of the jungle. The destruction they wrought is still being assessed, but it looks as if the Bornean rain forest is even more vulnerable than had been previously supposed.

The only beneficiaries of this disaster, which choked huge areas of Southeast Asia in a life-threatening smog, are oil-palm companies that set many of the fires and then smugly watched them go out of control, in order to gain for themselves new acres on which to plant their valuable crops.

In a hundred years human activities will have altered Borneo beyond recognition. The current changes have already severely affected the orangutan populations. Reports emerging from eastern Borneo tell of more than a hundred new orphan orang babies that have ended up in the villages, their parents killed by the fires or by poachers.

The inevitable loss of the Bornean ecosystem is a small part of a worldwide tidal wave of changes, not all of which will be entirely destructive. The next century will see far more changes than the last one, even though the last century has certainly been no slouch in the change department. A hundred years from now our culture and technology will be changed almost beyond recognition. We can only hope that it will give us the capability of undoing some of the thoughtless damage that we are now inflicting on our planet.

The evolution of orangutans has been slow, and they seem unable to adapt to environmental change to the same extent that we can— although they are being forced now to undergo at least some genetic changes. But because they were never caught up in the runaway-brain feedback loop, they will need our help in order to survive the destruction that we have caused.

If all this change is affecting the evolution of the hapless orangutans in their forest, it has a far greater impact on our own species. Humans have accelerated the pace of evolutionary change everywhere, and, at the forefront of that change, we are altering ourselves more rapidly than any other species.

This bald statement, of course, goes against conventional wisdom. Evolution, as we all know, is slow, and the rate of human-induced environmental change must surely be too swift for the glacial pace of evolution to keep up. But the conventional wisdom is wrong. Our gene pool, the huge collection of genes that together makes up all the genetic information in our species, is indeed responding to all this environmental change. In this book I intend to explore with you how this is happening.

PREVENTING SPECIATION

Orangutans, our third closest living relatives after chimpanzees and gorillas, are strikingly intelligent. They do not exhibit the same extensive social interactions and use of tools as chimpanzees, but on the nearby island of Sumatra they are more social than they are on Borneo and they have been observed to use and modify a number of types of tools in the wild. Indeed, a large project is now under way at Washington's National Zoo to determine whether orangs are as capable as chimpanzees of learning and comprehending language. So far, they seem to be scoring very high.

Orangs' intelligence extends to other areas as well. Known as the Houdinis of the primate world, they are notorious among zookeepers for their ability to escape from apparently secure cages by using a great variety of tools. Many airlines refuse to transport them, regardless of the precautions that have been taken to keep them confined.

Orangs are also becoming dangerously rare. It is estimated that about thirty thousand still thrive on Borneo, but only about a tenth as many have managed to survive on much more heavily impacted island of Sumatra. Human activity has already driven the orangs of Java and peninsular Malaysia to extinction.

Birute Galdikas made plain to me that, in spite of their confinement to the shrinking forests, we are forcing orangs into a new evolutionary direction. We are doing this by dramatically short-circuiting the evolutionary process that leads to the appearance of new species. Before humans appeared on the scene, the orangutans of Borneo and Sumatra, separated for millions of years, had been on their way to becoming different species. Now we are reversing that process, with consequences that cannot be predicted.

In spite of their separation, orangs from the two islands look very similar. In the zoological equivalent of a police lineup, even a trained primatologist would have a hard time distinguishing them. There are nonetheless invisible differences between the two populations, including small variations in their chromosomes that can be seen only under careful microscopic examination. DNA has been used to determine the approximate time at which the single ancestral group of orangs was split into two. Although there is a large range of error on this time estimate, the split appears to have taken place at least two million years ago.

These molecular investigations show that the two groups of orangs really are distinct, and as a result they have been categorized as separate subspecies. Biologists of the "splitter" persuasion have attempted to raise the two groups to full species status, though so far they have been unsuccessful.

Such quibbles might seem of interest only to taxonomists, that subgroup of biologists who spend their lives naming species. But a more important issue has to do with conservation: If two species of orangutan really exist, then this provides a powerful reason for preserving the habitats of both of them.

It is surprising that the orangs of the two islands have been separated for so long because the islands themselves have not always been separated geographically. When the Pleistocene series of ice ages began, about 1.64 million years ago, the sea level dropped and rose again repeatedly. Whenever the glaciers began to spread far to the north, dropping sea levels caused Borneo, Sumatra, and most of the other islands of the Indonesian archipelago to become joined together into a huge land mass. Orangs must have had ample opportunities to migrate among the islands during these episodes.

On the other hand, the land bridges that temporarily joined the islands may have been environmentally unfriendly places for the highly specialized orangutans. Or, even before the ice ages began, subtle behavioral differences may have accumulated between orangs in various parts of the great archipelago. These differences might have been enough to prevent any hybridization when the islands grew together and the animals were able to meet.

Any such behavioral differences that they have managed to evolve, however, can easily be overcome in captivity. Bornean and Sumatran orangs have mated readily in zoos and have produced a number of offspring that appear to be perfectly normal (although these offspring have not been tested for survival under fully wild conditions).

The zoo matings took place before the clear genetic differences between the Bornean and Sumatran subspecies were found. The discovery of these differences has thrown the zoos of the developed world into a state of confusion. In many but not all zoos, the two groups are now kept carefully separate. Some programs have been started to sterilize the hybrid offspring that have already been born, but in other cases the hybrids have simply been sold to less scrupulous zoos. Many other zoos, particularly the numerous private zoos in Asia, have made no effort to keep the two groups separate.

All this might not matter to the genetic integrity of the wild populations, were it not for the growing illegal trade in baby orangs. This trade is particularly rampant in Sumatra but is also growing in Borneo. Poachers kill the mothers that they track down in the forest, then take their babies to Java, where they sell them to unscrupulous dealers for about five hundred dollars each. In Semarang, in northern Java, I encountered a number of people who told me in detail about this ugly trade. Many of the infant orangs are sold to rich Taiwanese who keep them as house pets or put them in private zoos.

Cute baby orangs, like tiny kittens, inevitably grow up. Few households can withstand the rampages of a clever two-to-four-hundred-pound animal, particularly one that can bite through a human limb as if it were a wet noodle. Most house-pet orangs are killed when they become inconveniently large, but a few are returned to Java or Sumatra. Sometimes the babies are intercepted by the police during the kidnapping process and taken back to what the authorities suppose must be their native haunts. Galdikas, like other scientists at research stations on both Borneo and Sumatra, finds more and more of her time is consumed by the need to care for the growing numbers of orphaned orangs appearing on her doorstep.

Most of the orphans, accustomed to the plentiful food provided by humans, have tended to remain near Camp Leakey even after they have been released. But Galdikas has succeeded in returning a few to the forest. One, named Priscilla, was among the orangs that we observed swinging through the treetops about two kilometers from Camp Leakey. Priscilla's return was a brilliant success, for through binoculars we could see her tiny baby, Popeye, clinging tightly to her body. He had been born after his mother was returned to the wild.

None of the orangutan rehabilitation projects have had the resources to sort out the returned orangs genetically, to see which are from Borneo and which are from Sumatra. Some inadvertent gene flow has probably already taken place between the populations on the two islands. More is certain to occur.

One evening during my visit, as the rain poured down outside, Galdikas and I talked at some length about the probable consequences of this recent gene flow. The story is complicated. Both Borneo and Sumatra are huge islands, among the largest in the world, and the genetics of most of the orang populations that inhabit them has been little explored.

Galdikas suspects that gene flow might already have taken place among the islands, perhaps during the glacial maxima, long before the arrival of humans. The very orangs that she has been working with living as they do relatively close to Java, peninsular Malaysia, and Sumatra—might have been in the middle of this flow. A recent visitor to her camp, who had observed many Sumatran and Bornean orangs, concluded that her animals seemed to fall somewhere between the northeastern Bornean orangs and those of Sumatra. But this anecdotal observation was based on appearance and behavioral differences that are difficult to quantify.

Questions abound, and finding some of the answers may take years. How much gene flow has already taken place between the islands? How much will take place as orang populations are rescued by conservationists from soon-to-be-destroyed forests and transferred to other parts of the islands? The accidental gene flow that has already occurred, through the recent introduction of Sumatran orangs into Borneo and vice versa, is likely to be dwarfed by the gene flow caused by the ecological upheavals to come.

Any slight deleterious effects of the new gene flow, however, are probably the least of the orangs' current problems. The recent fires are an extreme manifestation of the invasion of Borneo by foreign lumber and plantation companies. These companies are accelerating their wholesale destruction of Borneo's coastal forests and are pushing inexorably farther inland. In another century any remaining orangs will be living in fragments of rain forest, perhaps covered by air-conditioned domes for the convenience of tourists. Any slight differences between the rainforest environments of Borneo and Sumatra will be dwarfed by the immense changes that are going to take place in the orangs' entire universe.

As the evening wore on, Galdikas and I speculated that, in view of the coming ecological disturbances, introducing some new genes into the Bornean and Sumatran populations might even be a good thing. In the days when the orang populations had adapted to slightly different environments on the two islands, such gene flow would almost certainly have been harmful. But during the present period of blindingly rapid change, it might now be advantageous. Gene flow, and the resulting new mixes of genes, might actually help orangs adapt to the few small altered remnants of their world that we will allot to them. In order to survive, orangs may need all the genes they can get. The more genetic variants their populations have, the more likely that adaptive gene combinations will arise and aid in their subsequent adaptation.

During my slogs through the humid leech-infested swamps, as I craned my neck to glimpse orangs in the sunlight and breezes of the high forest canopy far above, I reminded myself that these primates have been subjected to some of the same evolutionary forces that have shaped our own species, but that the outcome has been very different. We humans have, quite without meaning to, literally taken over and begun to redirect orang evolution, as we have with so many other species—including our own.

Like that of orangs, our ancestral lineage has been subdivided again and again. Sometimes this subdivision has resulted in the appearance of two or more separate species. Sometimes, as is now happening with orangs, the process has been aborted as two partially separated gene pools came together again. What happened when partially separated groups of our own ancestors were reunited?

Evolutionary theory suggests that such blending or fusion events, because they produce new combinations of genes, should help to accelerate evolution. Indeed, blending is perhaps the most powerful evolutionary process affecting us at present. Although no human groups are as genetically different as the Sumatran and Bornean orangs, some have been separated for 100,000 years and perhaps more. Throughout the world, however, different human races and groups are coming together. The consequences of this mixing are likely to be profound—and generally positive.

By the time I left Borneo, I had a much clearer idea why, in spite of their remarkable intelligence, orangutans did not follow human beings (or precede us) on an evolutionary feedback loop of the type that led to our own runaway brains. They had never been faced with the kind of sudden environmental changes that must have challenged our ancestors. The sadly battered state of the island itself vividly brought home to me how we are forcing the orangs' evolution in new and unpredictable directions.

What remained was the central question of how we humans are forcing our own evolution. The story of how we are affecting the Bornean orangutans forms just one small part of the answer to this question. Besides the orang story, other pieces that I have assembled in order to tell this tale include some recent fossil finds in a deep ancient cave in northern Spain, the way that people survive on the windswept reaches of the Tibetan plateau, a well-concealed evolutionary drama that is taking place in the hushed corridors of Whitehall, a superb adaptation of some of our primitive relatives to the rainforest of Africa's Côte d'Ivoire, and a glimpse of our frenetic future in the pell-mell world of extreme sports.

I hope you enjoy reading about the answer to this question as much as I have enjoyed putting it together, and that this book will leave you with a new appreciation of the power of evolution to change our species.

Many people have contributed to the book, including Juan Luis Arsuaga, Kurt Benirschke, Jack Bradbury, José M. Bermudez de Castro, Dan Bricker, Eudald Carbonell, Rusty Gage, Birute Galdikas, Tony Goldberg, Jean-Jacques Hublin, Michael Marmot, Richard Moxon, David Metzgar, Jon Singer, Ajit Varki, John West, Anne Marie Wills, Elizabeth Wills, David Woodruff, the members of the LOH study group, Helen deBolt, and the students of OSLEP. I would particularly like to thank Harvey Itano and Lorna Moore for detailed help. Ted Case and Pascal Gagneux read the whole book and made many valuable comments. As always, however, I am ready to take the fall for any errors.

Notes are sequestered at the end of the book, where they belong. Although I have tried to define technical terms when they first appear, a glossary is also provided to help the reader thread his or her way through the complicated bits.

PART I

The Many Faces of Natural Selection

ONE

❖

Authorities Disagree

Ever since Darwin, pundits—expert and otherwise—have made pronouncements on whether human beings are still evolving and what might become of us in the future. But these authority figures have historically disagreed enormously. The biggest battle has been fought about whether natural selection will continue to drive human biological evolution, or whether our species will instead degenerate. Surely, if the pressure of natural selection slackens, an accumulation of harmful mutations will sully our gene pool. These mutations, which are appearing now at much the same rate as they always have in the past, might soon overwhelm the weakened ability of natural selection to remove them.

Bald statements of this fear are less common now than they were before political correctness became a concern. In the 1880s Charles Darwin's cousin Francis Galton founded the Eugenics Society to encourage the selective breeding of highly intelligent people. In the 1920s and 1930s distinguished scientists such as geneticist H. J. Muller and evolutionist Julian Huxley lent their reputations to predicting the genetic destruction of our species. Huxley, in his 1948 book *Man in the Modern World*, made this grim prognosis:

> The net result is that many deleterious mutations can and
> do survive, and the tendency to degradation of the germ-
> plasm can manifest itself . . . Humanity will gradually

destroy itself from within, will decay in its very core and essence, if this slow and relentless process is not checked.

In 1950 Muller coined the term *genetic load* to express the burden of harmful mutations carried by our species, and he predicted dire consequences if nothing was done to remove them.

Some of this degenerative change is physical and has already taken place, anthropologist C. Loring Brace suggested in 1963:

> [S]ome of the major and formerly unexplained changes which have occurred in human evolution are the results of probable mutation effect. Reduction in the size of the teeth and face and of the skeletal and muscular systems may have been brought about by such a mechanism, as a result of changes in the principal human adaptation mechanism, culture.

Many authorities have claimed that mental degeneration has taken place and will get worse with time because of the accumulation of harmful mutations in the gene pool. They have cited a variety of studies showing how in developed societies people at the top of the socioeconomic ladder have fewer children than those near the bottom. The claim that this tendency will lead to degeneration has given rise to innumerable polemics about the inevitable destruction of our gene pool.

So concerned were these scientists about this perceived trend that they tended to ignore the political consequences of their statements. By the end of World War II, the question of whether our gene pool was accumulating harmful mutations, which perhaps had started out as a legitimate evolutionary inquiry, had become hopelessly tainted by ugly racist rhetoric and by the terrible deeds the Nazis carried out in the name of eugenics. The presumption of degeneracy itself was questioned, and many scientists who had a far less pessimistic view of our genetic future began to speak out.

Still, the theme of species degeneration sometimes bobs up in the mainstream. In 1997 writer Jerry Adler used the Millennium Notebook page of *Newsweek* to proclaim, "For Humans, Evolution Ain't What It Used to Be." In his article he suggested that while telephones might "improve" in the future, we will not!

Adler was at least careful to couch his worries in general terms. But is there any biological evidence for degeneration? Not really. Consider the apparent physical degeneration that concerned anthropologist Brace. Peoples around the world do seem to have become slighter in build, perhaps beginning with the introduction of fire to cook food and accelerating after the agricultural revolution. Our long bones have become lighter, and even our teeth have become smaller. Brace measured the reduction in tooth size in many different populations over time and calculated that it has averaged about one percent per thousand years—although he found much variation from one population to another.

Such changes mean that we are less capable than our ancestors of surviving in a world in which brute strength and the crushing power of mighty teeth are all-important. But these changes are not necessarily degenerative. We have, on the contrary, every reason to suppose that present-day humans are more capable than those ancestors of surviving in a world in which food is plentiful and easy to gather and chew, and in which small bodies and teeth are advantageous. For example, brute strength may be incompatible with fine motor skills—perhaps it is possible to select for one but not for both. An environment that places a premium on fine motor skills might then select for the kinds of changes that Brace has measured.

Brace and others have also proposed that apparently "degenerative" changes are the result of a relaxation of natural selection and the consequent accumulation of harmful mutations. In order to fit the "degeneration" model, these mutations would have to make us lighter boned and generally smaller. Brace argued that if there had originally been selection for mutations that made teeth and bones large, then simply by chance subsequent mutations would be likely to make them smaller.

We know from studies of many animals and plants, however, that most mutations are harmful because they introduce noise into the system. Thus, while random mutations might well lead us to have funny-looking jaws and teeth, there is no obvious reason why they would make our jaws and teeth smaller.

If we have simply been accumulating random mutations, moreover, it seems surprising that our jaws and teeth would have become smaller at such a regular rate while at the same time remaining well formed.

We can learn something about such changes and how they happened by examining an extreme example, pituitary dwarfism. When this condition appears in otherwise normal families it can be traced to an underproduction of pituitary growth hormone. A number of drastic mutations can produce this kind of dwarfism, ranging from a defect in a gene that releases growth hormone to defects in genes that result in damage to the pituitary gland as a whole. The harmful consequences of these mutations often extend far beyond dwarfism, so that the mutations do not persist for long in the population.

Pygmies of the African Efe tribe, on the other hand, have been tiny for millennia, and have thrived in their rainforest home. They have the shortest stature of any African pygmies, yet their pituitary function is completely normal. Their dwarfism stems from the fact that the cells in their bodies show a low responsiveness to the growth hormone, which their pituitaries manufacture in normal quantities. Even though the exact mutation has not yet been found, it is already apparent that it is far less damaging than the rare familial dwarfism genes that appear in small numbers throughout our species.

Selection for slight build has also played a role in the domestication of animals. The size of teeth and jaws in domesticated cats has been reduced compared with their wild relatives (as has, incidentally, their brain size). We would not, however, dream of attributing this reduction to the accumulation of mutational noise—cats are too elegant for that. Far more likely, we have selected for cats that do less damage to us and our belongings. In so doing we have presumably drawn on variation that was already present in the wild population from which the cats originally came. Any harmful mutations that appeared during this process would have been selected against—the genes that have produced these changes in cats will probably turn out to be more like the genes that produced short stature in the Efe.

The selective forces that have driven us in the direction of a mild kind of dwarfism are probably the result of a trade-off that gives an advantage to increased or altered brain power and fine motor skills despite decreased bodily strength. These changes, like the changes that took place in the cat population, have drawn on genetic variation that was already present in our ancestors. Not all of our distant ancestors, after all, were huge and burly with large teeth.

Another possible driving force behind such an evolutionary trend, as Darwin himself was the first to realize, might be sexual selection. If standards of desirability in a mate have shifted from sheer size and raw power to something a little bit less intimidating, the resulting sexual selection might explain the decrease in robustness of our species over the last few dozen millennia.

Oddly enough, these possibilities have seldom been suggested as explanations. Later in the book we will look at some suggestive evidence that such trade-offs have taken place, reinforcing the argument that selective pressures, acting on preexisting variation, are far more likely to have produced such a shift than the degeneration of our gene pool through the accumulation of harmful mutations.

If degenerative change is a shaky explanation for our decreased robustness, it is even less persuasive as an explanation for a decrease in our intellectual powers. The ugly, popular, and potentially racist view is that, because the most intellectually fit among us are not reproducing in sufficient numbers, our entire species is becoming stupider and more degenerate. Although this idea can be traced back to the ancient Greeks, it was first articulated compellingly by Francis Galton.

Galton, like most of his contemporaries, was what one might term a genteel racist. He took it for granted that some races were superior and some were inferior. Yet if he is able, from his vantage in heaven, to observe what has happened to his ideas in the ninety years since his death, I feel sure that he must be truly horrified. Even today, racists and supremacists of every stripe (the stripes are usually but not always pale ones) continue to cite Galton to support their belief in the intellectual degeneration of people unlike themselves. They persist even in the face of strong evidence that our species seems, if anything, to be getting smarter over time. In Chapter 12 we will examine the evidence that runs counter to the intellectual degeneracy argument, and the probable reasons for this remarkable and encouraging trend.

Whether or not we are actually getting stupider or more physically degenerate over time, it still seems obvious to most people that natural selection has pretty well stopped, bringing our physical evolution to a halt. This is because our environment is now less severe and dangerous than it once was. At the same time, as many observers have pointed out, this relaxed physical environment has freed us to do other things, so that our cultural evolution has sped up. To be sure, cultural evolution is not the same as physical evolution: a human population may acquire an elaborate culture without any changes in its genes. So it is quite possible that physical evolution has come to a halt at the same time as cultural evolution has roared ahead.

In 1963 the prominent evolutionary biologist Ernst Mayr of Harvard University wrote in his magisterial book *Animal Species and Evolution:*

> To be sure, there may have been an improvement of the brain [over the last 100,000 years] without an enlargement of cranial capacity but there is no real evidence of this. Something must have happened to weaken the selective pressure drastically. We cannot escape the conclusion that man's evolution towards manness suddenly came to a halt. . . . *The social structure of contemporary society no longer awards superiority with reproductive success* [my italics].

In 1991 English geneticist Steve Jones took to the op-ed pages of *The New York Times* to ask, "Is Evolution Over?"

> [M]ost babies born now survive until they themselves have babies, so that existence is less of a struggle than it was. Natural selection involves inherited differences in the chance of surviving that struggle, and as most of us do survive nowadays until we have passed on our genes, the strength of selection has decreased. . . . It may even be that we are near the end of our evolutionary road, that we have got as close to utopia as we ever will.

Mayr and Jones both fell into the trap of assuming that just because natural selection is not as obvious as it once was, it must therefore be coming to a halt. But as we will see, selection is still acting on us, even though the emphasis is now shifting away from obvious physical selective factors to ones that are more psychological.

A slightly less dogmatic view was espoused by physiologist and evolutionist Jared Diamond, in a 1989 article for *Discover* magazine:

After the Great Leap Forward [approximately 35,000 years ago], cultural development no longer depended on genetic change. Despite negligible changes in our anatomy, there has been far more cultural evolution in the past 35,000 years than in the millions of years before.

Anthropologist Richard Klein, in a 1992 article in *Evolutionary Anthropology*, made the same point.

[A] major transformation in human behavior occurred about 40,000 years ago. Prior to this time, human form and human behavior evolved hand-in-hand, *very slowly over long time intervals* [my italics]. Afterward, evolution of the human form all but ceased, while the evolution of human behavior or, perhaps more precisely, the evolution of culture, accelerated dramatically.

Quite correctly, both authors make a clear distinction between changes in culture and evolutionary changes in the genes that influence how our brains and our bodies work. They essentially rule out the latter.

Klein is right about the very long period over which we have evolved. As he also correctly points out, many important changes took place in our ancestors during the whole span of four million or so years for which we have a fossil record, while relatively few have taken place recently. But both he and Jones are, I think, wrong about the pace of evolution. In fact, our evolution has been blindingly fast compared with that of our closest relatives, all during that four-million-year period.

Nor is there hard evidence that our physical evolution has slowed down. Even though relatively little change seems to have taken place during the last forty thousand years, this span of time represents only one percent of our fossil record. The argument of Diamond and Klein would be more convincing if we had a really plentiful fossil record of our ancestors throughout our history, so that we could measure just how much change actually took place during various forty-thousand-year periods in the more distant past. But we do not—the gaps in the record are often hundreds of thousands or even millions of years long. And as they themselves admit, some physical change has even taken place during the last forty-thousand years. Was it more or less than the one percent we would expect if the change were spread evenly over our four-million-year fossil history? We simply do not know, and so we really cannot say whether our physical evolution has recently slowed down. Indeed, we cannot rule out the possibility that it has sped up!

By contrast, another group of writers are confident that we are still evolving—although they tend to waffle a bit about how much and in what direction. They do, however, emphasize far more than the others we have quoted the extent to which cultural pressures can, like any other sort of natural selection, act to modify us.

Roger Lewin, in his 1993 book *Human Evolution*, raises the possibility of an invisible kind of selection, one that might affect something other than our bodies:

> Why . . . would social complexity have taken 90,000 years to manifest itself after the origin of anatomically modern humans? One possibility, of course, is that a subtle intellectual evolutionary change may have occurred rather recently in human history, one that does not manifest itself physically.

At the end of his controversial 1975 book *Sociobiology*, which examined how behaviors evolved, Edward O. Wilson expresses this viewpoint even more clearly:

> Starting about 10,000 years ago agriculture was invented and spread, populations increased enormously in density, and the primitive hunter-gatherer bands gave way locally to the relentless growth of tribes, chiefdoms and states. . . . There is no reason to believe that during this final spurt there has been a cessation in the evolution of either mental capacity or the predilection toward special social behaviors. The theory of population genetics and experiments on other organisms show that substantial changes can occur in the span of less than 100 generations, which for man reaches back only to the time of the Roman Empire.

A similar view comes from David Hamburg, again from 1975; but he goes a step further, pointing out that cultural change is in fact continually moving the goal posts. The selective pressures that result from cultural change are themselves changing:

> The poignant dilemma is that ways of fostering survival, self-esteem, close human relationships, and meaningful group membership for hundreds, thousands or even millions of years now often turn out to be ineffective or even dangerous in the new world which man has suddenly created. Some of the old ways are still useful, others are not. They will have to be sorted out, and sorted out soon.

This small sampling of opinion shows that, over the last few decades, our view of the evolutionary future of humans has shifted. The discovery of mutations, and the demonstration in the 1920s that they can be induced by environmental factors, led to many dire predictions that our gene pool would become damaged. The apparent slackening of the pressures of natural selection would seem to have put the brakes on our future adaptive evolution, leaving these harmful mutations to accumulate unchecked. But as Hamburg suggests, selection pressures have actually shifted from the visible to the invisible, from forces that test the resources of our bodies to forces that test the resources of our minds.

WHITHER *HOMO SAPIENS?*

Lord Ronald . . . flung himself upon his horse and rode madly off in all directions.

STEPHEN LEACOCK, *Gertrude the Governess*

Assuming we are evolving, what might we be evolving *into?* Suppose, as I have suggested, that human beings are becoming better adapted to the changing environment that we are creating for ourselves. This is, after all, what evolution would be expected to do. Assuming that this trend continues, what might eventually happen to us?

Science fiction writers commonly conjure up speculative futures in which they suppose that we will evolve over the next few centuries into hairless, dome-headed, spindle-shanked creatures of great intelligence.

Scientists, too, have made such predictions. In 1967 anthropologist C. Loring Brace drew a tongue-in-cheek cartoon of just such a creature (see Figure 1–1). How this remote descendant of ours might possibly hold his head up is unclear. But Brace went the science fiction writers one better, for he hung around his neck something that looks remarkably like a Sony Walkman. Not bad for 1967!

Science fiction writers, of course, make a living by extrapolating trends. The fossil record and other sources of evidence tell us that the bodies of our ancestors have indeed become weaker and less hairy, while their brains have become larger. So these science fiction scenarios certainly seem like logical extensions of trends that we have observed. But will these trends really continue? Does anything about the process of evolution suggest that they will? Or are other things more important—at least in the short term?

To understand what might happen to us, we have to understand something about evolution: Selection-driven evolution does not necessarily always go in a particular direction. Even if, over the next few centuries or millennia, some directional changes do take place, they are likely to be dwarfed by other changes that will take place in response not to overall trends in our environment but rather to its rapid diversification.

As people today find more effective ways to shield themselves from the extremes of the natural world, it might be supposed that our environment is becoming less rather than more diverse. It is fashionable to decry the global Americanization of world culture, for example, which enables us to stay in a Holiday Inn in Marrakesh or to buy a Big Mac in Uzbekistan. Yet most inhabitants of these places have no interest in either activity, and for those who do, their environment has actually become more diverse, not less.

Around the world the number of types of occupations has exploded, as have the opportunities that have opened to people of both sexes and all races. Children of Uzbek sheepherders are now traders on the Tashkent stock exchange, perhaps dealing in mutton futures. Children of Tuareg nomads thrive on city life in Marrakesh (and one of them may even be the manager of that Holiday Inn).

A second thing to remember is that all this change is being paralleled by changes in the human gene pool. These genetic changes are taking place more rapidly than we might think.

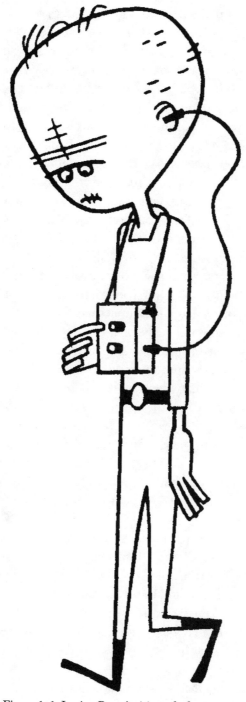

Figure 1–1. Loring Brace's vision of a future man.

Most of us suppose that evolution is always a very slow process. A favorable mutation arises somewhere in a population and then spreads, followed by the eventual appearance of another favorable mutation, its subsequent spread, and so on. Since most newly arisen mutations either have little effect or are unfavorable, for every favorable mutation that does appear and manage to spread, many others must arise and then disappear. The whole process would seem to be achingly slow—most of the substantial evolutionary change that we can follow from the fossil record seems to take millions or tens of millions of years.

If our gene pool responds so slowly even to our rapidly changing environment, the argument runs, then our evolution could not be getting faster. Even if it is going on at the same rate as it did in the past, the argument would be that we are likely to see little result from it in the space of a few centuries or millennia.

Yet surprisingly rapid short-term evolution can take place. In Africa's Lake Victoria, which has appeared and disappeared repeatedly in the past, many new and distinct species of fish have arisen over just the last few thousand years. The current collection of fish species, diverse in their behaviors and appearance, has evolved during the ten thousand years since the lake was completely dry. So closely related are they still that when the lake becomes murky because of human pollution, they lose the visual cues that allow them to distinguish mates of the same species. Hybrid fish are now turning up in increasing numbers in the cloudy waters of the once-pristine lake.

Even in the absence of any new mutations, evolution can happen quickly. Dramatic changes in a population can occur simply through a shift in the frequencies of the alternative forms of genes known as *alleles*.

A simple example will make this plain. Suppose some characteristic of our species is controlled primarily by the activities of four different genes. The characteristic might be something obvious, like skin or hair color, or it might be something more subtle, like the ability to perceive spatial relationships or to imitate a particular pitch of sound. Often there may be a trade-off: sheer muscle strength, as I suggested earlier, may be incompatible with the ability to perform small precise movements.

Suppose further that two alternative alleles of each gene are present in our species. Depending on our parentage, for each gene we could have inherited two copies of one allele, two copies of the other, or one of each. Since there are three possible combinations of the two alleles of each of these genes, there are $(3 \times 3 \times 3 \times 3)$ or $3^4 = 81$ possible combinations of these genes, or *genotypes*, in our population.

We will assume further that the different alleles of each of these genes interact in an additive way. Let us call the two allelic forms of each of these genes 0 and 1. Suppose that allele 0 of each gene results in more of one characteristic, while allele 1 results in more of another. Then, if the character governed a trade-off such as the one between strength and fine motor skills, a person with genotype 00,00,00,00 might be very strong but clumsy, while a person with genotype 11,11,11,11 might be relatively weak but have a high level of the skills needed to chip an elegant stone arrowhead—or repair a watch.

Whether a given population will carry all eighty-one of these combinations, and in what frequency, will depend on the population's history. Suppose that, through chance or selection, one human group ends up with a predominance of 0 alleles. There would be many strong people in that group, and few if any expert watchmakers. If another group ends up with a predominance of 1 alleles, the situation would be reversed. These two groups would be, for this characteristic, very different from each other, and there might be relatively few people in either population with an "average" mixture of the two characters.

Two such populations, each made up of ten thousand people, are shown in Figure 1–2. In the first population the frequency of the 0 alleles for each of the four genes is eighty percent, while in the second the frequency of each 0 allele is only twenty percent. To make things fair, a great deal of environmental "noise" has been added to this genetic picture: the genetic contribution to this characteristic is assumed to be only fifty percent, while the other fifty percent has been supplied entirely at random.

Even with all this random environmental noise, you can see that the two populations are clearly different from each other. Almost nobody in the first population scores an 8, and almost nobody in the second population scores a 0.

Now suppose that these two formerly isolated populations come together, intermarry, and after a few generations become thoroughly mixed. The result, seen in Figure 1–3, is very different. After mixing, the majority of the population is made up of people with average

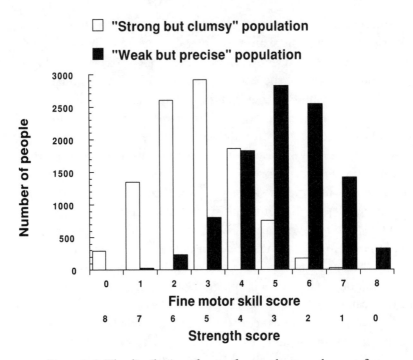

Figure 1–2. The distribution of scores for muscle strength versus fine motor skills in two different populations. In the first population, the frequencies of alleles of four genes that contribute to these abilities are all at 0.2, and in the second they are all at 0.8. Note that, because environmental fluctuations have been added to the picture, it is not possible to determine an individual's genotype from his or her score or *phenotype*.

attainment, and there are some people at each extreme. Unlike the first two populations, all eighty-one gene combinations are found in this mixed population.

The ability of this mixed population to move in different evolutionary directions has been enhanced, for it has a wider range of *phenotypes* than either of the original populations. (The phenotype is simply the appearance and abilities of an organism—what you see, in other words—and is the result of a complex interaction between the genes and the environment.) The people who make up the population will on average be slightly more variable, since the proportion of genes that are heterozygous in an individual will have risen slightly.

It is important to remember that none of this evolutionary change has been brought about through the appearance of new mutations.

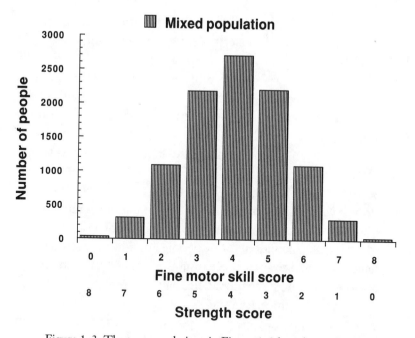

Figure 1–3. The two populations in Figure 1–2 have been allowed to interbreed. This population has a wider range of phenotypes, and therefore a greater evolutionary potential, than either of the earlier populations.

All that has happened is a shift in the frequencies of existing alleles. Of course, the alleles of the four different genes did originally arise by mutation in the past, but none had yet managed to take over entirely in either of the original populations.

Moreover, only four genes are required to produce this clear separation between the original populations. In the human population there are thousands of genes that have different alleles and that influence many different characters. No wonder that all of us, with the exception of identical twins, are so different from each other— and that our species has so much potential for future evolution.

This little exercise was designed to show you that any population can have an enormous evolutionary potential, so long as it happens to be plentifully endowed with genetic variation in the form of a wide assortment of alleles. It is a myth that species need new mutations in order to evolve. At least in the short term, old mutations that have already accumulated will do quite nicely, though if evolution is to continue for long periods new mutations will eventually become necessary.

Nor do we need to wait for one allele to replace another in order to increase this evolutionary potential further. Suppose that, as the result of a mutation, an allele arises that increases fine motor skills at the cost of a further reduction in brute strength. We will call this allele 2. Through chance or selection it rises in frequency until alleles 0, 1, and 2 are all equally frequent. Then a new allele arises at one of the other genes, but this allele enhances strength at the expense of fine motor skills. This allele, too, increases in frequency. What would be the effect on the population?

As you can see in Figure 1–4, the average of the population has changed very little as a result of these new mutations—most of the people still have average characteristics. But the range of abilities in the population has increased. A few people are now very strong, and a few have extreme dexterity. Increasing the range of variation in the population, even though the new alleles have not replaced the others, has increased its evolutionary potential still further.

If both such alleles appear, the population has not changed in any obvious direction, but its genetic variation has increased. Such

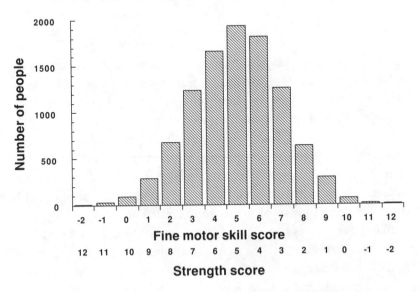

Figure 1–4. The effect of adding a new motor skill allele of one of the four genes and a new strength allele of another. The range of characteristics in the population has increased.

changes are nonetheless evolutionary changes, for they alter the composition of the population's gene pool and increase its range. I contend that such changes make up at least as important a part of the evolutionary process as any obvious directional selection for bigger heads or less hairy bodies.

Now, at the end of the twentieth century, different human groups, carrying different mixes of alleles, are blending together with unparalleled speed in almost every part of the planet. The rate of this blending can only increase in the future. As a result, the evolutionary potential of our species is growing enormously.

SORTING OUT THE POTENTIAL

What will happen to all this evolutionary potential, and whether it will really matter to our survival, is a different question. Even if the genetic variation in our species is increasing, selection is not necessarily acting on it. Without selection, regardless of whether our gene pool is changing, most of us would conclude that we are not truly evolving. It is my task to convince you that much of this variation really does matter.

To begin, we must remember that the history of our species over the last few million years, *and continuing down to the present time*, is one of pell-mell evolutionary change.

Much of this selection has been for sheer diversity. No other species on the planet is as behaviorally diverse as our own. Composers and concert musicians, brilliant mathematicians and engineers, linguists, superb athletes who have the ability to perform physical tasks that are at least the equal of what the most athletic of our primate relatives can accomplish—all these and more show that we have a growing ability both to create and to take advantage of a cultural and environmental diversity that would have been unimaginable a few millennia or even centuries ago.

The evolutionary changes that we are undergoing today are not taking us in one single direction, as those science fiction writers have tended to suppose. Instead, like Stephen Leacock's hero, we are galloping in many directions at once. It is now possible to construct testable models for how all this genetic diversity has been maintained and increased over time. We will see that selection for diversity, particularly behavioral diversity, has been going on for a long time, and that it has been greatly enhanced by the complexity of our current societies.

WHERE IS THE SELECTION?

In 1962 the eminent geneticist Theodosius Dobzhansky published a book called *Mankind Evolving*, which addressed the question of whether we are still evolving and concluded that we are indeed. As a beginning graduate student, I was most impressed by the book. I was delighted soon afterward to meet the author and have the opportunity to work with him.

In retrospect, Dobzhansky timed his book rather badly. Shortly after he wrote it, a flurry of amazing discoveries about our fossil record began, a flurry that has continued and accelerated down to the present and that has totally changed how we think about human prehistory. And decades later new genetic research about our species exploded onto the scene. Nonetheless, many of the matters that concerned him turn out to be timeless themes that we will also address in this book. Perhaps we will push a little closer to some answers.

Much of the last part of Dobzhansky's book directly addressed the question that we are wrestling with in this chapter: Just how much is natural selection likely to be affecting our species at the present time? Here, oddly, he found himself on rather uncertain ground. He could not point to a great many clear-cut examples of such selection.

To be sure, he emphasized that a great variety of diseases still exert selective pressures on human beings, particularly in underdeveloped countries, and that in those regions malnutrition continues to play an important selective role. But he had trouble finding examples of selective pressures acting on people in the industrialized world. He did his best, telling a few tales of selective pressures that technology has introduced. For example, he noted that people who suffer from a genetic disease called porphyria are highly sensitive to certain barbiturates, and that a number of them died after treatment with these otherwise relatively benign drugs. But such examples of selection, as he would have been the first to admit, are unlikely to have had much of an impact on our gene pool. Of course, he wrote at a time when our species' impact on the planet was only just beginning to be perceived—the environmental disaster stories that assault us daily were then unknown.

Dobzhansky did make the important but usually unappreciated point that the gene pool of our species is not simply the result of selection *against* things, such as susceptibility to diseases. Nor is it entirely the result of selection *in favor of* things, such as upright stance, bigger brains, and cleverer hands. Rather, our gene pool is a complex and dynamic collection of alleles that rise and fall in frequency and reshuffle with each generation in ever-shifting alliances. Because there is a diversity of environments, there will be selection for diversity in the gene pool.

Dobzhansky also pointed out that human beings clearly differ from all other species. Because of our intelligence, we have achieved a unique status in the world of living organisms. For millennia we have had the power to make sweeping changes in entire ecosystems—and even to destroy them utterly, as has happened in many parts of the Middle East since the invention of agriculture.

In spite of these insights, it was hard for him to find a convincing answer to the central question: Does our current success at altering the world around us, and our parallel ability to insulate ourselves from the natural world, provide us with immunity to the process of natural selection? Are harmful genes still being removed from our gene pool at the same rate as before? Is our evolutionary potential still increasing?

Today, despite a few contrarian views that echo Dobzhansky's, the received wisdom is that these things are not happening. Our recent conquest of many diseases is the example that is most often trotted out to explain the presumptive disappearance of natural selection.

Infectious diseases have certainly been powerful sources of selection on our species in the past. Yet over the last few centuries, and particularly in the last century, mortality from these diseases has plummeted in the developed world. Americans now live nearly twice as long on average as their great-grandparents did. In much of the world, any healthy child born now has an excellent chance of growing up and having children in turn. The obvious conclusion is that our culture, especially our new technologies, is shielding us from the kinds of traumatic adventures that our millions of ancestors experienced.

Although the argument has some justice, it is less forceful than it first appears. There is no doubt that the obvious impact of natural selection has diminished during our very recent history, particularly for those of us who are lucky enough to live in the developed world—although new pressures have also been introduced, such as allergies to newly invented substances. But the apparent diminution of selection raises two questions.

First, diminished from what? If selection operated on us in the past with particular vigor, then even after its effects have lessened, a great deal of it may still be going on.

The second and more complex question is, what does natural selection really mean in our species? In the industrialized world, people happen not to be dying in large numbers in obvious and premature ways, but this does not mean that selection has necessarily ceased to operate. Natural selection has many more subtle routes by which it can act on the variation-filled gene pool of a complicated species like ourselves.

TWO

❖

Natural Selection Can Be Subtle

Though Nature, red in tooth and claw,
With ravine shrieked against his creed . . .
ALFRED LORD TENNYSON, *In Memoriam* (1850)

Natural selection sometimes moves in mysterious ways, its wonders to perform. To understand how it works, we must carefully examine the concept itself.

INVISIBLE SELECTION

It is a fallacy to assume that the effects of natural selection are always visible. Sometimes they are quite invisible. To take just one example: A couple chooses not to have children. This conscious choice, whatever its reason, ensures that their genes will be lost to the gene pool. The effect is the same for a couple that is unable to bring a baby to full term. The absence of a baby who was never conceived, or who was aborted very early in development, can have a greater effect on the gene pool of our species than the untimely death of an adult who has already reproduced.

There are two general reasons why a baby does not reach term. The first is that at some point, usually early in development, something goes severely wrong with the developing embryo. Estimates vary, but it would appear that between twenty and thirty percent of fertilized eggs never complete development. Some of these failures happen late enough in pregnancy to be detected, but most occur so early that the mother may not even have suspected that she was pregnant.

But in the overwhelming majority of cases the reason an embryo dies is a gross genetic abnormality—missing or extra chromosomes, or even the presence of an entire extra chromosome set. This last severe defect accounts for ten percent of the spontaneous abortions that occur in the first trimester.

The influence of the environment can also be profound. During the 1984 Bhopal disaster in India, huge quantities of the deadly gas methyl isocyanate were released into the atmosphere. In the succeeding weeks and months the numbers of reported spontaneous abortions among the survivors who had been exposed went up almost fourfold, to a quarter of the total number of pregnancies. (The actual rate is certain to have been much higher than the reported rate.) Many of these extra abortions could be traced to chromosome abnormalities, apparently induced by the gas exposure.

Bhopal was an extreme case, but it was not unusual. Most of us are lucky enough not to live in India's ironically named "Golden Corridor," where industrial disasters continue to happen even after Bhopal. But we are all affected to some degree by the rise of chemical waste in our environment.

These chemicals may be applying their own selective pressure on our population, preventing the births of babies who would otherwise have developed normally. We simply do not know the extent of this pressure, or whether it is increasing or decreasing, but we do know that the numbers of accidental large-scale releases of dangerous chemicals are going up dramatically in many parts of the world. What are the evolutionary effects of these mini-Bhopals, and of all the everyday, less spectacular releases of contaminants?

Declining male sperm counts have been much in the news lately. It has been suggested that chemicals produced by the plastics industry that resemble female hormones might be to blame, and such chemicals can indeed cause sperm counts to decline in rats. But we may be worrying about a nonexistent phenomenon. The studies that claim to show declining counts in humans have been strongly critized in evey aspect, and it is now unclear whether the declines are real or merely statistical artifacts.

However, men who already have low sperm counts may be particularly susceptible to the effects of environmental contamination. A recent study of a group of such men in Athens, Greece, showed a significant decline over a seventeen-year period. If sperm count really is falling among such susceptible groups, then this is certainly a brand-new selective pressure that is being brought to bear on at least part of the human population.

The second reason that a baby might not be born is a relatively new one in human societies: Couples can now choose, more than ever before, whether to have a baby at all. This factor adds an entirely new and extremely complex psychological dimension to the influences shaping our gene pool. The genes that contribute to psychological makeup of a population are fair game for selection, and as we will see in Chapter 5, the effects and impact of such genes may well be increasing.

Not all of the invisible selection that alters our gene pool is selection against physically or psychologically harmful genes, or against sperm, eggs, or embryos that have been damaged by the environment. Selection can act to preserve and even to enhance the genetic variation within a population, or to increase its genetic capabilities in other ways.

Such alterations are often missed. When scientists follow the evolution of various plant and animal species over time, they generally measure changes in the average value of some character that is of interest to them. The increase in the human brain size is an obvious example. But far fewer studies follow changes in the variation of a character around its mean. Yet such changes are an integral part of biological evolution. It would be most important to know, for example, how much selection there has been for new alleles of genes that influence our psychological makeup, as the psychological landscape of our societies has become more complex. Although we do not yet know the answer to this question, we are on the point of being able to find out.

IMITATIONS OF NATURAL SELECTION

Invisible natural selection is likely to be changing our gene pool quite rapidly. But its effects will soon be supplemented dramatically by our growing expertise in genetic technology. We will soon have the power to make changes in our genes that resemble the changes caused by natural selection.

The first such changes will be straightforward "patches" that cure defective genes. Early attempts at such gene therapy have met with a notable lack of success, but a new generation of modified viruses that can carry undamaged genes into damaged cells is starting to become available. At the moment gene therapists are concentrating their efforts on putting functioning genes into affected tissues—for example, inserting a normal CFTR gene into lung cells of cystic fibrosis patients or a normal hemoglobin gene into the blood-forming tissue in people with sickle cell anemia. A strict moratorium has been imposed on making such changes in the cells that produce eggs and sperm and that can therefore pass the altered genes on to the next generation.

But once gene-delivery systems become more sophisticated, and more precise gene surgery becomes possible, the moratorium will undoubtedly be abandoned. W. French Anderson, one of the pioneers of gene therapy, is already arguing that research on germ-line therapy should be permitted. Eventually, as the technology gets better, his argument is likely to prevail.

If it can be done neatly and without damaging any other genes in the process, patching a sickle cell anemia gene so that the cured rather than the damaged gene will be passed on to the next generation seems like a very sensible thing to do. But it will be a short step from patching genes that clearly cause specific diseases, to patching genes that cause less obviously harmful conditions, and from there to modifying genes in such a way as to enhance our abilities. Genes that contribute to obesity, small stature, and other stigmatizing conditions are already fair game. As long ago as 1984, transgenic mice carrying a rat growth hormone gene were produced; they reached twice the size of their littermates. And, recently, mice with a defective gene that results in obesity and diabetes have been given successful gene therapy in the form of a virus carrying the gene.

These manipulations in rats and mice point the way to such manipulations in humans. Genes discovered in these animals are almost certain to have human equivalents. Human genes that increase muscle strength and even mental abilities will soon be discovered, and totally new modifications of these genes will quickly be invented. Success stories involving such genes will soon sweep aside any moral scruples people might have about altering our gene pool.

It is well to remember two things, however. First, such manipulations are very different from the natural process of evolution. Many of the changes that genetic surgeons carry out never have been tested by natural selection. There is nothing to prevent some of those changes from reducing a patient's long-term chances of survival or reproduction, just as the current craze for tongue studs and for nose, navel, and nipple rings increases the chance of infection.

What we consciously do to ourselves will always be, to one degree or another, in conflict with the often-invisible and hard-to-comprehend processes of natural selection. But we are sure to do it anyway. Many athletes, for example, currently use dangerous steroids to increase body mass. More than half of competitive athletes interviewed have said that even if the steroids were absolutely certain to cause them to die young, they would still use them, because winning is so important to them.

Second, the human gene pool is a very big place. With 5.5 billion people and 200,000 genes per person, the genes that are changed by deliberate intervention will simply be a drop in the bucket. Genetic changes driven by natural selection will continue to dwarf the efforts of even the most indefatigable gene therapists, at least for the foreseeable future. But chemical means of enhancing our bodies and our minds will also become commonplace in the future, and as we will see, these too can have an evolutionary effect.

Human beings have already been carrying out such gene manipulations on domesticated plants and animals, in some cases for millennia. Most of these changes were accomplished simply by artificial selection. The history of such efforts is surprisingly long: recent evidence shows that our ancestors domesticated dogs on two separate occasions, tens of thousands and perhaps more than a hundred thousand years ago.

The results of artificial selection have often been grotesque. When ears of corn and wheat are left unharvested in the field, they simply fall to the ground near the parent plant and rot as they germinate. If milk cows are left unmilked for a few days, they will die of burst udders. Cattle are now being bred for looks—but their straight rear legs, which give them a pleasingly square "cow" shape, can actually make it difficult for them to walk. Artificially bred tom turkeys can be so massive, they are unable to climb atop females and impregnate them. In order to do their job, the poor things need to be lifted up by amused farmhands.

Such animals and plants have been bred, not for characters that would aid in their survival in a state of nature, but for characters that people find useful. For those of us who shop too extravagantly at the new genetic mall, the results are likely to be just as grotesque, even disastrous. Clearly we are not yet wise enough to be able to imitate natural selection in all its aspects (and we may never be).

And, just to remind you how inextricably natural and artificial selection are intermixed: the people who succumb to the blandishments of this dazzling array of genetic "fixes" will not be a random sample from our gene pool. It seems likely that one result will be selection for genes that contribute to caution, and against genes that contribute to reckless behavior!

GENETIC EVOLUTION VERSUS CULTURAL EVOLUTION—A FALSE DICHOTOMY

The changes that we have wreaked on domestic animals and plants have generally damaged their ability to survive in the wild. But what about the things we have done to ourselves? There is no doubt that we have much reduced our own ability to survive under natural conditions. Our remote ancestors such as the Australopithecines, and our closer relatives such as *Homo habilis* and *Homo erectus*, lived during times of great and continual danger, threatened by animal predators, starvation, disease, and tribal warfare. The selection that acted on them was omnipresent and fierce. We, on the other hand, are largely protected from such severities and are able to survive and reproduce without hindrance. Our cultural prowess has helped us to modify virtually every part of the planet's surface and has enabled a substantial fraction of us to live long, disease-free, enriched, and stimulating lives.

As a result, we have become very well adapted to our cultural milieu—but remarkably ill adapted to the natural world. This shift has been going on for a long time. Even hunter-gatherers who live in the warm and abundant tropics today need at least a modest assortment of artifacts in order to survive. Most of us need a lot more than that.

Just think of all the things we must take along in order to spend a week in even as relatively friendly a natural environment as California's Sierra Nevada mountains—a region that famed outdoorsman John Muir called the Gentle Wilderness. Some of the artifacts we may take are fripperies, like freeze-dried food and perhaps Nintendo games. But others are brand-new necessities, such as filters for removing dangerous pathogens like *Giardia* from the waters of streams that used to be pure before legions of backpackers polluted them. Muir himself, without the aid of fripperies, could wander off for months into those same mountains and manage very well, but he still required boots, clothing, a poncho, a knife, a gun, fire-making equipment, and so on. And to keep himself in touch with the human world, he took along books to read and notebooks to write in.

The selection we have imposed on ourselves has been, in many ways, as artificial as the selection we have imposed on domesticated animals and plants. Strip our culture away, and all of us—except perhaps a few dedicated and well-practiced survivalists—would likely perish. We have gotten ourselves into this situation, and the human gene pool has changed greatly as a result.

Because of this triumph of culture over nature, one might think, it is now our culture and not our genes that will primarily determine what will become of our descendants.

If true, this transition would mark an enormous shift in human history. It would mean a complete change in the influences that operate on us, with predominantly biological forces being replaced by predominantly cultural ones. We will see, however, that it is not true. The powerful effects of our culture have, if anything, accelerated our biological evolution.

This acceleration began very early. Our remote ancestors lived far more intimately with the savage world of nature than we do, yet they were remarkably successful creatures. They were already able, through their social activities, to diminish the more obvious rigors of natural selection. Yet their evolution did not slow—rather it accelerated. This seeming paradox provides clues to the ways in which selection is currently acting on us.

Genes and culture have been intertwined for a long time. The appearance of crude stone tools in the fossil record marks the first real indication of culture as we understand it. In their earliest manifestations these tools were simply stones that had been partially shaped to produce a sharp cutting edge at one end. Detective work has shown that they really were the creations of our hominid ancestors. For example, they are rarely found in isolation. Sometimes they are discovered in such large numbers that our ancestors must have come together in groups over long periods to make and use them. They also often have a different composition from unmodified rocks found in the same area, showing that they were carried by their makers over substantial distances.

The time span over which our ancestors can be shown to have made stone tools is continually being lengthened. Tools dated to 2.5 million years ago have been discovered in southern Ethiopia, in Israel, and in Pakistan. Even at this early stage in their history, our tool-making ancestors were peripatetic and adventurous, traveling beyond their African homeland far sooner than had been previously supposed.

Perhaps most surprisingly, the tools in these old deposits were already relatively sophisticated. Discoveries of even older stone tools are therefore sure to be made sooner or later. Even though 2.5 million years ago seems extremely ancient, we have yet to uncover the true beginnings of human tool-making.

The bones of our ancestors are anything but rarities in the fossil record, and if abundance can be considered a measure of success, then our ancestors were remarkably successful. Even before tools appeared in the fossil record, social interactions of some kind, however primitive, must have contributed to their achievements.

Toolmaking is only part of this story of accelerating evolutionary change, however. By 2.5 million years ago, for example, our Australopithecine ancestors had already been walking upright for at least a million years—though perhaps not with quite the same panache as modern humans. And there is certainly no evidence that the invention of tools halted their subsequent physical evolution, for afterward our ancestors continued to undergo many other dramatic changes in brain and body.

If culture did not halt our genetic evolution back then, why must we suppose that it has done so now?

REVEALING OUR EVOLUTIONARY POTENTIAL

We will soon be able to modify our own bodies and even our own genes more dramatically than we have modified those of domesticated animals and plants. But even in this brave new world, the things that we are able to do to ourselves will still be dictated by our evolutionary histories and the capabilities that we have acquired through natural selection.

Take brain size. The size of the human brain does not seem to be entirely governed by some innate mechanism that tells it to stop growing at a certain point. Rather, its final size, at least in part, is dictated by the rate of maturation of the skull in which the brain is imprisoned.

One baby out of every three thousand is born with a defect that causes one or more of the sutures of the skull to close too soon. This premature cranial synostosis distorts the shape of the head and puts great pressure on the brain as it grows. The problem can be corrected surgically, by preventing the suture from closing until after the brain has matured. As a result of this drastic intervention, the disfigurement and mental retardation that would otherwise result can be ameliorated and in a good many cases prevented entirely.

Cranial synostosis brings the growth of the brain to a premature halt by a combination of increased intracranial pressure and decreased blood supply. My colleague Kurt Benirschke recently pointed out to me that we simply do not know the constraints that a normally developing skull puts on the size of the brain. Suppose that, in some nightmare surgeon-dominated world of the future, it becomes common practice to perform craniotomies on normal babies. How large would their brains grow, once the barriers were removed? What effect, if any, would this have on their intelligence?

Of course, this change would be surgical rather than evolutionary. In the absence of further surgical intervention, the children of these deliberately altered people would have brains of a normal size. A true evolutionary change would take place only if larger brains were to prove advantageous to our survival in the world of the future, selecting for genes that delay the closure of skull sutures (among many others) so that the change would become a permanent part of our species's gene pool.

Benirschke's brain example may seem far-fetched, but few of us in the developed world now get through life without surgical interventions of some sort. Sometimes we owe our continued existence to them.

These interventions are often dictated by our evolutionary history. For example, one consequence of our upright posture is that our teeth are crowded toward the backs of our jaws. Most readers of this book have probably had some or all of their wisdom teeth surgically removed, either to make more room for the others or because they were perversely growing into the teeth next to them.

The collision of evolutionary forces that has led to this particular problem must have been dramatic. Dental surgery is a recent invention, going back at the most a few thousand years. During much of this time the methods that were used to treat tooth problems, while often effective, do not bear thinking about. A Roman skull has recently been found in which an artificial iron tooth had been inserted into the upper jaw—where it seems to have functioned quite well for some years before the patient died, apparently of an unrelated cause. A few years ago I watched a traveling dentist set up temporary shop in the public square of a village in Morocco's Anti-Atlas Mountains. As dentists in that part of the world had done for centuries, he extracted teeth from his surprisingly docile patients with the aid of a fearful set of rusty forceps and a large mallet and chisel. But whenever the wisdom teeth of our ancestors became impacted, the results must have been disastrous, for such primitive surgery could not have corrected that problem. The advantages of upright posture must have been overwhelming indeed, since they outweighed the negative consequences of such severe and often life-threatening dental problems.

The difference between the natural and the artificial worlds is not completely clear-cut, however. If the surgeon's knife can increase brain size—and perhaps even intelligence—by a relatively simple operation, then our brains already possess a potential that is normally masked by the opposing process of skull maturation. This potential, even though almost never realized under ordinary circumstances, may be an evolved property of the genes that control the development of the brain itself.

We are not likely to be alone in possessing this potential: other organisms must exhibit a similar property. Suppose that an operation were carried out on the skull sutures of a chimpanzee baby, so as to stretch out the period of maturation of its brain. The operation would almost certainly not result in a chimpanzee with a human level of intelligence, but it is not beyond possibility that its intelligence might be measurably increased.

Such a potential would have developed over a long period of time. Suppose that the brains of our distant mammalian or reptilian ancestors had sometimes had the opportunity to grow to unusual sizes, perhaps because those ancestral creatures happened to carry mutant genes that delayed the development of their skulls. If this genetic accident resulted in an increase in brain power, then genes that delayed skull maturation might have been selected for. So would genes that helped to adapt the brains to their new roomier homes. Thus, in the future world that we have just envisioned, the surgeon's knife would simply be exposing an evolutionary potential that has been revealed many times before, albeit under different circumstances, during our long evolutionary history.

The same thing applies to brain biochemistry. As I write, "smart" drugs are becoming the rage. Many of us are dosing ourselves with extracts of the leaves of the living fossil tree *Ginkgo biloba*, or ingesting huge quantities of rather scary biochemicals such as acetyl carnitine that are supposed to increase the activities of the energy-producing mitochondria in our cells. These substances, it is claimed, will make us think better—and indeed there is some evidence that ginkgo extract actually slows down the rate of onset of Alzheimer's disease.

These "smart" drugs may or may not work on people with normally functioning brains—there is as yet no evidence one way or the other. But the public's desire for a pill that will make one think better is so palpable that when drugs that really do work—as is inevitable—are found, the market for them will doubtless be huge. Already memory-enhancing drugs are in the testing stage at a number of pharmaceutical companies, and even more remarkable advances are on the way. A mutation that enhances long-term memory and decreases training time has recently been found in the fruit fly, and the hunt is on for the equivalent gene in humans.

In yet another variant of this brave new world, children could be given drugs that would enhance their musical or mathematical or artistic abilities. These enhancements, however, will be possible only because the genes that these drugs affect are already there—all our brains have these capabilities built in as a result of our evolutionary history. It seems likely that these drugs will have untoward and unexpected side effects, particularly at first, and that disasters will be as numerous as success stories. But we will learn from our mistakes and get better at such manipulations. (In Chapter 14 I will explore some of these incredible scenarios and their likely evolutionary consequences.)

HOW THE ENVIRONMENT CAN ENHANCE
GENETIC POTENTIAL

Even after the dust settles and we get really expert at altering and enhancing our brain function, it will be impossible to divorce the environmental changes we make from evolutionary changes that are happening at the same time. The more we know about altering our brains, the more we will force biological evolution, because genes and environment always work together.

Selection cannot act on genetic differences if those differences are not in some fashion made manifest in the bodies or minds of the people who carry the genes. As we learn more about our genes, we are discovering more and more situations in which genetic effects results from our modification of the environment.

For example, some 300 million people worldwide, primarily males of African or Mediterranean ancestry, carry a sex-linked condition called G6PD-deficiency. This condition protects against malaria, but if there is no malaria in the environment, the carriers of the gene suffer no untoward effects—that is, unless they are incautious enough to eat broad beans or breathe the fumes of mothballs! The result of these apparently innocent activities can be a life-threatening anemia. This condition is revealed only when the carriers of the gene are exposed to this particular narrow range of environments, some of which—like the mothballs—are of our own making.

Another pervasive environmental factor, which we have also generated ourselves, is smoking. A lengthening list of genes is being discovered that, in their carriers, enhance the deleterious effects of smoking. Smoking is a stupid thing to do, but for these people it is particularly stupid—the genes they carry would normally predispose them to emphysema or lung cancer only slightly, but smoking greatly enhances their risk.

Our livers are responsible for breaking down and rendering harmless all sorts of unpleasant chemicals in the environment, and they do so by manufacturing many different enzymes specific for the destruction of various classes of such chemicals. A liver enzyme called paraoxonase, for example, is manufactured by a gene that comes in a wide variety of alleles.

One of these alleles, found in about a third of Asians and ten percent of Caucasians, turns out to protect against the effects of organophosphate pesticides. These chemicals, vanishingly rare in our past but now essential to twentieth-century agriculture, are currently responsible for three million cases of severe poisoning and 220,000 deaths worldwide every year. Another allele of the gene gives little protection against these chemicals but protects against nerve gases such as sarin. Both of these alleles, of course, appeared in the population long before either kind of chemical became common. Whatever their original function, it is apparent that these alleles are going to play a much larger role in our adaptation to the horrendous world of chemicals that awaits us.

Organophosphate pesticides will be common in that world. We can only hope that disasters like the release of sarin into the Tokyo subway system by the Aum Shinrikyo cult in 1995 will not also become common.

The selective pressures that originally caused these paraoxonase alleles to be retained at substantial frequencies in the human population still remain a mystery. The presence of the allele that protects against sarin is particularly baffling, because it has recently been shown, in a study carried out in Japan, to predispose diabetics to heart attacks. Nonetheless, whatever their original function, our recent manipulations of the environment have made these alleles far more important than they were in previous generations.

The rule that genes can be selected for or against only if their effects are made visible also applies to genes that affect brain function. Brain-enhancing chemicals will be part of the environment of the future. Different people will, because of their genes, respond to some of these chemicals far more readily than others. Suppose we find ways to artificially enhance such mental differences between people, and this in turn has an effect on whether they reproduce. Then such manipulations are likely to accelerate the process of natural selection as it acts on these differences. Even as these soon-to-be-invented biochemical manipulations extend our capabilities, they will magnify the effects of natural selection.

It is not impossible to imagine the following scenario. Suppose that somebody with a potential talent is unable, through a variety of environmental circumstances, to realize that talent. Discouraged and feeling useless, he or she might turn instead to a high-risk life and suffer an early death before reproducing. But if, through some kind of biochemical or other environmental enhancement, that individual instead achieves early success, it might put his or her life on a totally different and far more fortunate track. We all know of cases in which a chance encounter with a dedicated teacher changes a pupil's career. But the teacher, in order to work that magic, must realize that the talent is there and waiting to be nurtured. Perhaps "smart" pills will help this process. Stranger things have happened.

We must always remember, as we move into a world in which surgery and biochemistry offer new ways to explore the potential of our bodies and minds, that this potential is itself a product of evolution. Realizing that potential is something that our current highly challenging environment can do very well, and it seems likely that this trend will continue.

In the middle of the eighteenth century, Thomas Gray was inspired to write a poem by some anonymous graves he found in a country churchyard. In his elegy he described the graves' occupants as being like flowers born to blush unseen and waste their sweetness on the desert air. At the end of the twentieth century, entirely because of enhanced genotype-environment interactions, far more such human flowers should be able to flourish and contribute to our cultural richness.

The converse is true as well. People who are ill adapted to the modern world will be less likely to have children—not because their own survival is directly threatened, but because of psychological pressures that they would not have encountered a century or two ago. Much of what is happening to our species at the moment is being driven by these relatively invisible psychological factors. Their effects are not as obvious as the effects of predation, starvation, and disease, but they are nonetheless potent. And at the moment, because our physical environment is not as severe as it once was, these psychological pressures are becoming more important.

As I explore with you these interwoven evolutionary processes, the important theme that will emerge is that evolutionary change can happen in many different ways. There is much more to evolution than selection for big brains and hairless bodies.

FEAR OF NATURAL SELECTION

In addition to the claim that natural selection has ceased, another enormous objection has been raised to the idea that we are continuing to undergo biological evolution. This objection emerges from a fundamental misunderstanding of a new and important idea in human affairs: that of social equality.

During the last three centuries, the wondrous Enlightenment notion that human beings are all, at least potentially, socially equal has come ever closer to realization. It has liberated an increasing fraction of us from the bondage of superstition, and from servitude to kings and tyrants. Throughout much of the developed world, societies have arisen that are predicated on this concept. If we are all socially equal, then it is self-evident that we should all be provided with an equal opportunity to make ourselves into something greater.

But the idea of social equality would seem to be in direct collision with the idea of genetic determinism. How can we construct a socially equal society if we are prisoners of our genes? How can our new and hard-won social freedoms liberate us from the genetic tyranny of natural selection?

The proponents of social equality justifiably fear that if nature is really more important than nurture, so that the differences between people are genetic rather than cultural, then our characteristics and fates must be unavoidably written in our DNA. If we are prisoners of our genes, then even though we have managed to overthrow the human tyrants of the past, we have simply exchanged one sort of bondage for another.

Because these new bonds are built into our bodies and minds, they will be far more difficult to break. And it is only a short and easy step from the conclusion that there are large genetic differences among individuals to the inference that there is a biological basis for race and class prejudice.

This concern is important. Bitter experience has taught us that the idea that we are genetically unequal, as it is usually presented, is incompatible with the idea that we are socially equal. Distressingly, when a scientist or other authority figure points out that people can be divided by some criterion into superior and inferior groups, it immediately gives the imprimatur of science to those who would maintain social inequalities.

The latest manifestation of this destructive effect is a book by Richard Herrnstein and Charles Murray called *The Bell Curve*. Utterly uninformed by any evolutionary perspective, the authors present at length a great deal of evidence for the existence of genetic inequalities among individuals. Some of this evidence is convincing, some much less so.

At the end, triumphantly waving their collection of factoids, they attempt to use them to justify and reinforce the social inequalities that have shaped our society. After all, they point out, these inequalities certainly exist. Our genes are in the process of shaping our society, and they produce an intellectual caste system in which those at the top are the natural and inevitable overlords of those at the bottom.

I find the Herrnstein and Murray approach abhorrent, because they take the undeniable fact that humans differ genetically and turn it to base and short-term political ends, writing off much of our species as unsalvageable.

In Chapter 13 I will deal with the issue of nature and nurture from the perspective of an evolutionary biologist. The evidence is overwhelming that, because of the genes they carry, some people are better at certain tasks than others. But what does this fact actually mean, in terms of its immediate impact on society as well as its long-term evolutionary effects? Not, we will discover, what Herrnstein and Murray thought that it means at all.

The famous bell curve of IQ scores is a far more complex phenomenon than is generally supposed. A dramatic increase in certain kinds of IQ scores, called the Flynn effect, has taken place over the last few decades in industrialized societies. This change is not evolutionary—it is happening far too quickly for that—but growing evidence suggests both a genetic and an environmental basis for this phenomenon.

The environment plays a larger role in such things as intelligence than had been supposed just a few years ago. It is now becoming apparent that it is impossible to establish a hierarchy of intelligence. Not only is intelligence manifested in many different ways, as psychologist Howard Gardner has pointed out, but the genes that contribute to it are constantly shifting and recombining from one generation to the next. This means that even though you may be particularly smart—and insufferably smug about it to boot—there is no way of predicting how intelligent, or in what fashion, your grandchildren will be.

The only constant in this bewildering story is selection for a diversity of intellectual capacities. The genes that contribute to these capacities are part of the rich tapestry of our evolutionary heritage. Follow the threads for even a few generations, and you will find that everybody contributes to the tapestry and that it is impossible to decide which of the threads is most important.

IQ and other human abilities are labile and easily modified, and it is the height of cruel foolishness to write off any group of humans. All of us have untapped abilities and carry important genes essential to the future survival of our species. It is the diversity of our gene pool as a whole that holds the key to our future evolution, not some particular set of genes that form the basis for an intellectual elite.

THE FUTURE

In Chapter 14 I will gaze, with some trepidation, into my crystal ball and speculate about what is likely to happen to our species in the future—provided of course that our story is not brought to a close by the sudden arrival of an asteroid. Particularly in this era of swift cultural change, prognostications are nearly always rendered irrelevant by events. Nonetheless, I will try my hand at examining the potential that is held out by the astonishing efflorescence of science that has taken place during the last two centuries and that will continue in the centuries to come.

The looming crises that confront our species are primarily of our own making. The terrible fires in Southeast Asia are only the latest in a series of environmental wake-up calls. The realization is growing that our current population size cannot be sustained without severe damage to the planet. War and disease are unlikely to control our population—we will have to do it ourselves, by modifying our behavior. There is no way we can begin to reduce our population over the next century or two without selection playing some kind of role, and I predict that this selection will primarily be at the behavioral level.

On the other hand, as far as the increased diversity of our species is concerned, the sky is no longer the limit. If life is found elsewhere in the solar system or beyond, some of us will surely visit and eventually colonize those places. As we spread to worlds that are very different from our own, the consequences for our evolution will be at least as profound as when our remote ancestors first ventured out of Africa.

Will people living on other planets evolve into new species? Given sufficient isolation from the rest of us, we have no reason to suppose otherwise. Indeed, challenged by very different environments, they will probably become new species much more quickly than the millions of years that were required to engender the timid beginnings of speciation in the orangutans of Borneo and Sumatra.

In short, the powers of natural selection that Darwin was the first to understand will certainly continue to shape our species. Here it is our task to try to glimpse the accelerating ways in which biological and cultural evolution will reinforce each other in the future, and to understand how this mutual interaction will allow us to survive the evolutionary challenges that we will face as we begin our spread through the universe.

THREE

◈

Living at the Edge of Space

*Those who sweat get frostbite easily, but I never sweat
when I am climbing. . . . Sometimes it has been suggested
that I have "three lungs" because I have so little trouble
at great heights. At this I laugh with my two mouths.
But I think it is perhaps true that I am more adapted
to heights than most men; that I was born not only in,
but for the mountains. I climb with rhythm, and it is a
natural thing for me. My hands, even in warm weather,
are usually cold, and doctors have told me that my
heartbeat is quite slow. . . . On a recent tour of India,
with the heat and the crowds, I became more sick than I
have ever been in my life on a mountain.*

TENZING NORGAY, *Tiger of the Snows* (1955)

Even though we have yet to visit other planets, some of us have spread
into environments that are almost as alien, with profound evolutionary
consequences. Think of these migrations as a kind of practice run for
our future attempts to colonize Mars or the planets of a nearby star.

The world has four roofs. One is in East Africa and encompasses
the highlands of Ethiopia; our ancestors have lived there for millions
of years. A second, which was colonized by humans only in recent historical times, is the Colorado Plateau. A third—far higher than either
of these and inhabited for perhaps ten thousand years—is the high
Andes, extending along much of the western rim of South America.

But the roof of the world that everybody thinks of first is the Tibetan plateau. This region encompasses some 800,000 square miles, an area more than twice as large as the high Andes. The plateau has been pushed up by a slow, grinding, but irresistible collision between two great pieces of the Earth's crust: the tectonic plate that carries the Indian subcontinent, and the immense mainland of Asia itself.

As India has plowed into Asia's southern rim during the course of the last fifty million years, the most obvious result of this collision has been the Himalayas. These mountains, however, form only the southernmost part of a massive pile of ancient seabed material that has been shoved ahead of the Indian plate, like snow in front of some vast snowplow. Squeezed into accordion pleats by the force of the collision, this material has been formed into range after range of peaks. Each spring fierce rivers wear away the rocks of the high-altitude valleys between the peaks, where in spite of everything a little life manages to cling.

The nearest I have come to this roof of the world was a visit to the desert town of Dunhuang in Gansu Province of western China. Dunhaung is located one hundred kilometers from the Tibetan Plateau's northern edge, squarely at a junction of the Great Silk Road. People have lived there for a long time. The famous Buddhist caves at nearby Mogao have yielded, among other treasures, the earliest known printed book, the famous *Diamond Sutra*, which can be dated from clues in the text to A.D. 868.* But these are by no means the first traces of human occupation. In the nearby Taklimakan Desert, Chinese and European archaeologists have found more than a hundred well-preserved mummies of a Caucasian people who lived in the area four thousand years ago. The origin and fate of these people, so different in appearance and culture from the present-day inhabitants, is a complete mystery.

The immediate countryside around Dunhuang is relatively flat, invaded in places by rolling sand dunes from the desert to the south. Beyond the desert is a range of six-thousand-meter peaks known as the Denghe Nanshan, which stand out vividly whenever the air is clear. Even from so far away, these immense mountains dominate the landscape. The sky to the south of them is pale with reflected snow from range after range of still mightier mountains. And it is there that some

---

* Earlier printed books are known, in particular books from Korea that probably date from before A.D. 750, but the *Diamond Sutra* is the first printed text that can be dated with precision.

members of our species who have migrated to the edge of space can be found.

A MYSTERIOUS HISTORY

We know little about the history of human penetration of the Tibetan plateau. Perhaps a Shangri-La or two is still hidden somewhere among those remote peaks and valleys. Whether there is or not, people have dwelt in this most extreme of lands since long before written records. The severity of their lives has, until recently, been counterbalanced by their remoteness, which has made them relatively safe from invaders.

Among the diverse inhabitants of the plateau, the Sherpas of Nepal are the people who have interacted the most with the outside world. They are not native to Nepal, but migrated there during the seventeenth century from southern Tibet. Sherpas are brilliant mountaineers and make up an essential part of any climbing expedition in the Himalayas. Their mountaineering skills and ability to carry heavy loads at high altitudes have made such expeditions possible and have enabled numerous well-heeled Manhattanites to stagger to the tops of the world's highest mountains. The Sherpa Tenzing Norgay was the second person to stand on the top of Everest, courteously dropping a few feet behind Edmund Hillary during the final scramble to the summit.

The Sherpas and other indigenous peoples of the plateau have adapted remarkably to the need to live their entire lives at around four thousand meters, where each lungful of air has only a third as many molecules of oxygen as at sea level. But has this adaptation been a truly evolutionary one? Do the Sherpas actually carry genes, or combinations of genes, that are different from those carried by people who live at sea level? If so, then in the course of acquiring these genes, they have undergone a true evolutionary change.

This question is difficult to answer because it is hard to distinguish between genetic adaptations and those that are merely physiological. Any human body has the capacity to adapt fairly rapidly to extreme conditions. At high altitudes the number of red blood corpuscles in our blood can rise substantially, and over time even our lung capacity can increase somewhat. Further, any of us can soften the impact of the cold simply by bundling up warmly, by not exposing our extremities more than necessary, and so on. Are the Sherpas really different genetically from the rest of us? Or do they have these physiological adaptations simply because they have lived their entire lives, beginning at the moment of conception, at high altitudes?

The theory of evolution predicts that any population that lives under such extreme conditions for many generations should become better and better adapted; the poorly adapted members of the population will die or fail to reproduce. In order to see whether natural selection has altered the gene pool of the Sherpas, we would like to have at least a rough idea of how long their ancestors have lived under these conditions. The longer this period has been, the greater the likelihood that selection has been able to change their gene pools.

We cannot be very precise about this estimate, because uncertainties abound. Very little genetic information about Tibetan populations is available, and the human fossil record in this region ranges from sparse to nonexistent. The Tibetans have a strong genetic and physical resemblance to the peoples who currently inhabit the steppes to the north of the plateau, in the area around Dunhuang and elsewhere. That region is likely to be the primary origin of the peoples of Tibet.

But if the origin of the Tibetans is shrouded in mystery, so is the origin of their ancestors. Those puzzling Caucasians of four thousand years ago, so different from the present-day peoples of the northern steppes, are unlikely to have been the first arrivals in the area. Indeed, the first arrivals on the steppes, and perhaps even the first to venture into the Tibetan highlands, might not even have been members of our own species.

Early hominids,* in particular our immediate forerunner *Homo erectus*, likely inhabited Central Asia for a very long time. Although fossil hominid sites in Central Asia are few and far between, two different very ancient sites have been discovered that (very loosely) bracket the Tibetan Plateau. Fragmentary remains of *Homo erectus* from both of these sites have been dated to almost two million years ago. One of the sites is far to the west of Tibet, in the Caucasus Mountains; the other is located to its east, in the upper valley of the Yangtze.

---

* Hominids include our own species and our very closest relatives, all of which are extinct—the Australopithecines, *Homo erectus*, the Neandertals, and so on. Hominoids, on the other hand, are the result of a wider cast of the taxonomic net and include our closest living relatives—the chimpanzees, bonobos, gorillas and orangutans.

Such old sites have not yet been found in Central Asia. But if bands of *H. erectus* did trudge across the vast windy space that separates these two known sites, it hardly seems likely that they did so only once. Much more probably, during this immense span of time, they migrated back and forth many times across what would eventually become western China and Kirgiziya. In doing so, they would have anticipated Marco Polo and the other much later travelers belonging to our own species, for they would have been following the route of the Silk Road—long before there was any silk or presumably even any concept of a road.

Of course, since no fossil record of such migrations exists, all this is sheer speculation. We also have no evidence that tribes of *H. erectus* ventured into the mountains that they would have glimpsed far to the south of their migratory path. If they had done so, they would probably have taken the most accessible routes, from the northeast and the southeast, in order to avoid crossing the vast Gobi Desert. But we do have evidence that, about half a million years ago, they were living close to the plateau itself. The evidence comes from a discovery made in a cave near the village of Yuanmou, not far to the southeast of the plateau's edge between Chengdu and Kunming. The scattering of teeth and tools found there belonged to *H. erectus* and have been tentatively dated to 500,000–600,000 years ago, though they may be younger.

It may not be coincidental that these traces of *H. erectus* happen to have been found near a well-worn trading route into Tibet that has been used extensively during historic times. There is no doubt that this difficult but negotiable route leading to the high plateau would have been open to these predecessors of modern humans. I find it hard to imagine that some of them, impelled by curiosity or the need to escape from enemies, did not venture into the mountains.

If *H. erectus* reached high altitudes and stayed there for any time, they would have become adapted to the conditions there. Then when they were displaced—and perhaps driven to extinction—by later arrivals, those adaptations may have disappeared with them. But perhaps not. *

* I will do my best to ignore the persistent rumors of the yeti, or Abominable Snowman. This mysterious hairy creature figures in elaborate legends told by the Sherpas, who claim that it dines on fungi found at high altitude. Footprints and supposed samples of yeti hair have all, when carefully investigated, been traced to other animals. Still, one cannot help but wonder what the origins of the legend might be!

The much more recent migrations through this area of our own species, *Homo sapiens,* are also mostly shrouded in mystery. For centuries the plateau has lured traders who have entered through routes that snake over the towering passes and lead into Tibet proper. Up until the time of the Chinese Revolution, endless files of Chinese peasants, bent almost double under eighty-kilogram bales of tea leaves, toiled up from Chengdu into the southeastern part of the plateau, in order to slake the Tibetans' great thirst for tea. Nineteenth- and twentieth-century travelers tell of endless lines of porters stumbling up slopes that would challenge a trained mountaineer, numb and empty-faced because they were unable to withstand the cold and fatigue without the aid of opium.

But these traders left few records. Essentially nothing is known of Tibet's history prior to the seventh century A.D., although Paleolithic tools dated to about fifty thousand years ago have been found on the northern part of the plateau. So we have no idea when the earliest modern humans might have arrived, whether they found the plateau already inhabited, and what might have happened to any earlier inhabitants. But we do know that, in spite of the strong physical and genetic resemblance of the present-day Tibetans to the peoples who inhabit the steppes to the north, they did not come from that area alone. The physical resemblance does not extend to the languages they speak, which have an affinity with those of the Burmese, to the southeast. This discordance between appearance and language suggests that the peoples of the plateau have had many long-term cultural contacts with a variety of peoples who lived at lower altitudes, and that the Tibetans must be an amalgam of many different migrations.

Thus it seems reasonable to suppose that, along with new languages and trade goods, new genes have been repeatedly infused into the gene pool of the Tibetans. The genes could have come from the tribes of the northern steppes, from peoples who lived in the ever-more-populous river valleys to the east that would eventually become China, and from the diverse inhabitants of the foothills of the Himalayas. However, if the genes already carried by the earlier settlers that aided high-altitude survival were sufficiently advantageous, they might have remained in the gene pool and spread in numbers even in the face of an influx of new genes from the outside.

While we cannot identify precisely the period of years that the Tibetans have had to adapt to high altitude, it is certainly in the thousands and probably in the tens of thousands. If *H. erectus* really did leave some genes behind in Tibet before vanishing, there is a slim possibility that this adaptive period might even have extended over hundreds of thousands of years!

To get a clearer idea of the impact that this long period of adaptation has had on the Tibetans, we can compare them with lowlanders who have lived at high altitudes for only a matter of months or years. Although lowlanders such as the Han Chinese settlers tend to be much less well adapted, it can always be argued that this is simply because they were not born and brought up under such extreme conditions. If a lowlander couple were to migrate to the Tibetan plateau and have a baby, then by the time that baby grows up, it might be just as well adapted as the Tibetans who have lived there for generations.

We do not know if this is the case. The Han Chinese who now live in Lhasa and elsewhere in Tibet do so for only part of the year, and pregnant Chinese women almost always descend to lower altitudes to give birth. So it is difficult to attribute any differences between Tibetans and these lowlanders to a genetic cause.

One way to disentangle true genetic change from mere physiological adaptation would be to examine another group, who have lived at altitudes as high as the Tibetans but for fewer generations. If the evolutionary scenario is correct, these people should have adapted to their environment less thoroughly than the Tibetans, since there would have been less time for new genes and new combinations of genes to appear in their gene pool. Luckily, such a group can be found. They inhabit the second greatest roof of the world.

BEFORE THE INCAS

The human history of that other roof, the great cordillera of the Andes, has almost certainly been briefer than that of the Tibetan Plateau. The Andes were also formed by a collision of tectonic plates, though not as massive as the one that took place between India and Asia. Here the encounter was between unequals. As the thin oceanic plates of the eastern Pacific encountered the massive plate that makes up the western edge of the South American continent, they were forced to bend down, pushing up the coastal rim in the process. This relatively mild encounter has not crushed and folded the Amazon basin, which lies to the east of the collision zone. Instead, it formed the Andes, an immensely tall but relatively narrow range of mountains: the world's second tallest after the Himalayas.

The ancestors of the Quechua and Aymara Indians who inhabit the high valleys of the Andes must have made their way up into these regions a relatively short time ago, probably five or six thousand years. For one thing, human occupation in the South American mountains was dependent on the development of crops that could support them. Further, because of the shorter distances involved, the Andean altiplano was far more accessible to lowlanders than the huge and remote Tibetan Plateau. Throughout the history of the Inca empire, and undoubtedly in earlier times, there was much migration from low to high altitudes and back.

Nonetheless, during this short history, these peoples have managed astonishing things. Their accomplishments were not, as is generally supposed, all due to the Incas. The Inca empire did its best to foster this notion by carefully destroying any records of the achievements of earlier civilizations. But we know now that they were responsible only for putting some finishing touches on a vast network of roads and bridges that stretched from northern Ecuador to a point south of present-day Santiago in Chile, extending eastward into Bolivia and Argentina. This network, one of the most remarkable construction feats in the preindustrial world, was actually the result of the efforts of many different cultures.

Along these roads, in the days of the Inca empire, relays of specially trained runners called *chasqui* could carry verbal messages, or messages coded in a knotted string, from Quito to Cuzco in five days. For long stretches the *chasqui* had to run at altitudes well above four thousand meters.

The problems faced by the first colonizers of the Andean altiplano, like those that faced the first venturers into Tibet, were more challenging than a simple lack of oxygen. For most of the year, nighttime temperatures in the Andes fall below freezing. The extreme dryness of the air and the lowered vapor pressure suck moisture from the body fairly quickly. The blood literally thickens, and the brain is deprived of essential oxygen, leading to the blinding headaches, dizziness, and sleepiness—and sometimes the edema and internal bleeding—of mountain sickness.

When I arrived for the first time in Cuzco, the magnificent ancient capital of the Incas, I was advised by friends to drink as much liquid as I could manage for at least the first twenty-four hours. By following this good advice, I was able to avoid the worst of the altitude effects that often incapacitate unprepared tourists. But the first arrivals in these high desert valleys had no such warning. Their children were particularly at risk.

Life has always been severe in this region. Studies of mummified human remains from several different early cultures, ranging from 1300 B.C. to A.D. 1400, show that fifteen percent of the women had died in childbirth or immediately post-partum. Even today, among the Indians of remote areas in southern Peru, disease and the effects of the rigorous environment kill three hundred out of every thousand babies during the first year of life.

The Aymara and Quechua have developed ingenious methods for preserving the lives of their fragile children. The youngest babies are wrapped in layer after layer of cloth, rendering them immobile. This conserves heat, moisture, and the child's energy. The child in its bundle is then placed in a carrying cloth or *manta*. The amount of oxygen reaching its lungs is reduced by all the layers, but because it cannot move very much, any damage from anoxia is presumably minimized. The baby is kept tightly wrapped for the first three months, but as it grows the wrappings are gradually loosened.

How much selection for sheer survival must have taken place before this effective method of protection was invented? Cumulatively, it must have been very great. Before infant mortality was reduced to its present, albeit still brutally high, level, uncounted tiny bodies were buried or interred in the stone mausoleums on the cliffs surrounding the altiplano valleys. Not all these deaths had a genetic effect on the population, but the fraction that did, over the millennia, had the cumulative result of increasing the likelihood of survival of each generation.

It is striking and perhaps significant that Tibetan mothers swaddle their babies as well, but the babies' heads are kept freer of clothing so that they can breathe more easily. Does this mean that a Tibetan baby is better able to withstand the fierce high-altitude conditions than a baby in the altiplano? As we will see, there is evidence that it is.

A RUSH OF BLOOD

Mark Twain remarked that even though history never repeats itself, it does tend to rhyme. So does evolution. Selection for different genes and combinations of genes must have occurred in the two adventurous peoples who colonized the Tibetan plateau and the Andean altiplano. Although the Andeans have adapted less well, both groups have acquired the ability to live under these extreme conditions.

So far, scientists have made only limited comparative genetic investigations of these two groups. They have yet to track down any specific genetic changes that have contributed to their adaptation. But it is almost certainly only a matter of time before some are found.

Perhaps the most telling indication of true genetic differences between the two populations comes from an examination of the weights of their babies at birth. In virtually all parts of the world, high altitude has a strong and significant negative effect on birth weight, even when babies are carried to normal term. The effect is detectable even when the differences in altitude are relatively small. On average, for every thousand meters of altitude, the birth weight of full-term babies decreases by about one hundred grams.

This pattern turns out to hold true for the Andean Indians: their babies weigh about four hundred grams less than babies born at sea level—for the metrically challenged, this translates into a difference of a pound or so. Such a large reduction in birth weight must contribute to the high infant mortality. Babies of acclimatized parents weigh a little more than unacclimatized babies born at the same altitude, but not much.

This pattern emphatically does not hold true for Tibetan babies. In a notable exception to the worldwide trend, their birth weight at term is the same as that of babies born at sea level.

These high birth weights are not some special property of conditions on the Tibetan plateau. Chinese mothers who have lived in Lhasa throughout their pregnancies obey the birth-weight-altitude law. Even though they usually descend to lower altitudes to give birth, they have smaller babies than do Chinese mothers who live at sea level.

How do Tibetan mothers manage to nourish their babies so efficiently at high altitude? To find out, Lorna Moore of the University of Colorado visited Tibet a number of times from 1983 to 1992. Although recent political upheavals have now made further work difficult, during that window of opportunity Moore and her coworkers carried out a number of important experiments on Tibetan volunteers (see Figure 3–1). Using as controls some Han Chinese living in Lhasa who had spent long periods of time at high altitude, they found that, in vivid contrast to the Han women, Tibetan women excel at supplying oxygen to their babies during pregnancy.

Using Doppler ultrasound, it was possible to measure the rate of flow in the arteries that supply blood to the lungs and to the uterus. In most people low levels of oxygen cause the pulmonary arteries, which supply blood to the lungs, to constrict, which is exactly the reverse of what they should do. This turns out to be a reflex, an echo of the moment when we are born. As a fetus grows in the womb, the pulmonary arteries are small and little blood flows to the lungs, but at

Figure 3–1. The two smiling women on the left were photographed in the Lhasa Bharkor market. They come from Kham, in eastern Tibet. The pregnant woman on the right is having her tidal volume and oxygen use measured. Pictures courtesy of Lorna Moore.

the moment of birth oxygen rushes into the lungs and the pulmonary arteries expand. We are now cursed with the reverse of this reflex—when oxygen drops, our pulmonary arteries contract in a physiological memory of fetal life. Moore and her coworkers found that this does not happen in Tibetans—soon after birth, they lose that dangerous reflex.

The rate of blood flow in the uterine arteries of the Tibetan women is very high for a different reason—their arteries are unusually large. These two factors combined mean that Tibetan women retain high rates of flow under hypoxic conditions.

After birth, other unusual adaptations are exhibited by the Tibetan babies themselves. During the period shortly after birth, Moore found that Tibetan babies had ten percent more oxygen in their arterial blood than babies born to Han Chinese parents who had lived at the same altitude. This difference persisted and actually grew more pronounced during the first weeks of life. Right from birth the newborn Tibetan babies were able to extract more oxygen from the thin air that surrounded them.

THE BODY'S PERCEPTION OF THE
HIGH-ALTITUDE WORLD

The fact that Tibetan mothers can reverse the effects of high altitude and give birth to babies of normal weight, while Andean mothers cannot, is the most powerful argument for real genetic differences between the two populations. But Moore and her coworkers discovered other things about their physiologies that also undoubtedly have a genetic component. Although these phenomena are complex and interrelated, they show that Tibetans have by far the best set of adaptations to high altitude of any human group so far examined—including even the Tibetans' neighbors, the high-altitude peoples who live in Ladakh at the northern tip of India. The people of Ladakh, like those of the Andes, ventured to these extreme altitudes more recently than the people of Tibet.

Both the similarities and the differences are informative. The lungs of both Tibetans and Andean natives have more of the tiny thin-walled sacs called alveoli, in which the blood is brought in close contact with air. But such adaptations to high altitude life are to be expected—even lowlanders can eventually develop them.

Another obvious accommodation to high altitudes is shown by the Andeans: They have more red blood cells, and higher levels of hemoglobin, than lowlanders. Lowlanders, too, tend to increase the number of red cells in their blood as they become acclimatized to high-altitude life. Remarkably, however, Tibetans show neither of these adaptations. In fact, they have slightly fewer red blood cells, and slightly less hemoglobin, than people living at sea level. This is not due to an increase in blood volume, which is the same as that of altitude-adapted Han. Further, when Tibetans climb to even more extreme altitudes than those at which they normally live, the hemoglobin in their blood does not increase to the same extent as it does in the blood of lowlanders when they become adapted to the same altitude. Recently suggestions have been made that an allele among Tibetans may control the rate of saturation of hemoglobin with oxygen, though hard biochemical evidence is as yet lacking. Oddly, the statistical analysis suggests that all Tibetans do not have this allele.

In the Tibetans increased blood flow is enough to overcome their deficiency in red blood cells. And their thinner blood has other consequences, the most important of which is protection against high-altitude pulmonary edema.

This severe illness constitutes the greatest danger to adult survival. When lowlanders climb quickly to high altitudes, their arterial pressure shoots up, especially the pressure in the pulmonary arteries. In mountaineers and others under severe stress, fluid is sometimes actually forced out of the circulation and into the lungs.

More insidious is chronic mountain sickness, in which the increased pressure forces blood out of the capillary beds into the tissues. A great variety of symptoms can result, some taking years to develop, including bleeding under the skin and gastrointestinal hemorrhage. The symptoms of the disease are obvious even to the casual observer: the faces of the victims are a darkened mahoganylike color, stained with the blood that has become trapped.

This slowly developing disease is common in people who have lived at high altitudes for long periods, or even for many generations. Surprisingly, it often develops in Peruvian highlanders, who one might have thought would have evolved resistance.

Chronic mountain sickness is also widespread among the Han Chinese living in Lhasa, who make it worse by incessant smoking. But it is almost never seen among Tibetans, even those who smoke heavily. Their cheeks remain pink and healthy without the telltale broken vessels or dark engorgement suffered by the lowland Chinese.

Resistance to the effects of dehydration may be the key to resistance to mountain sickness. Recall the remark of Tenzing Norgay's that I quoted at the beginning of this chapter. Tibetans may lose less moisture from perspiration, although this has not been systematically studied. In addition, the fact that Tibetans have a relatively small number of red blood cells means that their blood should not be as affected by dehydration. Even when they do lose fluids through exertion at high altitude, their blood will retain its normal low viscosity for longer. This too has yet to be studied in detail.

Yet another Tibetan adaptation, almost invisible but very effective, has to do with the way their bodies perceive the high-altitude world. They respond differently to the stresses of living at the edge of space.

When people who have spent their lives at sea level are first transported to a high altitude, their rate of breathing does the reverse of what one might expect—it actually slows down. This lowers the amount of oxygen in the blood and increases the amount of carbon dioxide, which can worsen the symptoms of mountain sickness. With time, as these newcomers become acclimatized, their breathing rates return to normal levels. Remarkably, the Quechua of the Andes, even after many generations at high altitude, always show a pattern of slow breathing, even though they are presumably thoroughly acclimatized. This may help to explain why chronic mountain sickness is so prevalent among them. They show what seems to be an inappropriate physiological response to their extreme conditions.

If you expect the Tibetans to show a different response from the Andeans, you will not be disappointed. At high altitude Tibetans breathe at the same rate as acclimatized newcomers. They do this because their bodies are sensitive to changes that are largely invisible to the rest of us.

In the laboratory it is possible to manipulate the amount of oxygen in the air supply that an experimental subject is breathing through a mask, and to do so in such a way that the subject gets no hint of what is going on. When oxygen is lowered to about two-thirds of normal, Tibetans react very differently from either Quechua or lowlanders. Their heartbeat rises rapidly from seventy to a hundred beats a minute.

The Quechua show a much smaller response, and lowlanders show almost no change at all. It takes a further substantial reduction in oxygen before their hearts begin to respond.

The bodies of the Tibetans, it appears, are far better than those of the rest of us at measuring oxygen levels. Their ability to sense conditions on the edge of space has been extended by the process of evolution, in just the direction that is needed to enhance their survival.

THE COEVOLUTION OF HUMANS AND YAKS

Tibetans are not alone in their superb adaptation to the high-altitude world. They share many of these adaptations with yaks, the high-altitude cattle of the Tibetan plateau. Yaks, too, have large and unusually thin-walled pulmonary arteries (Figure 3–2). This means that, under hypoxic conditions, the hearts of both Tibetans and their cattle do not have to work as hard as the hearts of lowland humans or lowland cattle to achieve the same results.

Just when the yaks first came to occupy the high-altitude plateau of Tibet is as much of a puzzle as the history of the first human occupation of the area. We do know that yaks have been plentiful in Tibet for a long time, long enough for people to have invented all sorts of uses for them. They pull plows in the high-altitude fields and carry huge loads over passes as high as six thousand meters. One recent traveler described "their great lungs inhaling and exhaling in puffs like a blowing locomotive." Indeed, they are so thoroughly adapted to high altitude that they tend not to do well below about three thousand meters.

In Tibet it is not sensible to waste anything. Every part of each yak is utilized. The hide has many uses, from clothing to coverings for the boats in which the Tibetans navigate the high lakes. The long and massive horns have been ground into powder for medication and have even been employed as building materials—in Lhasa, crushed yak horn is often used to reinforce masonry, and intact horns are inserted at intervals along the resulting walls for decoration. Yak oil is essential for lamps, and the butter is used to flavor tea, which the Tibetans drink in immense quantities to counteract the severe liquid loss at high altitude. Slabs of yak butter are often carved into intricate sculptures. So honored is the yak that two stuffed carcasses hang from

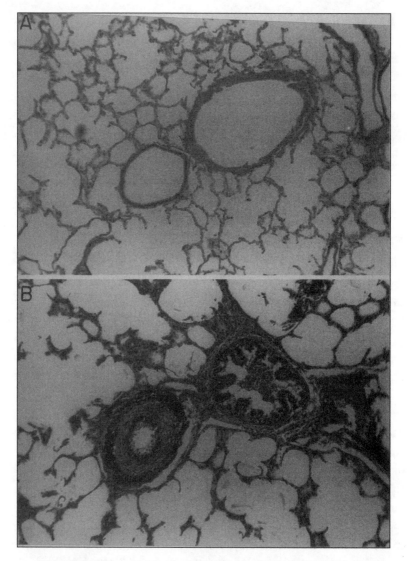

Figure 3–2. Cross-section of yak (top) and lowland calf pulmonary arteries (bottom) from Kurt Stenmark. In each case, the artery on the left is associated with an air passage on the right. Notice how thin-walled both the artery and its associated air passage are in the yak.

the ceiling of the great monastery that overlooks the remote town of Gyantse, southwest of Lhasa. Perhaps they are a relic of the time before Buddhism when the Tibetans worshipped a pantheon that included yak-headed gods.

Their long coarse hair gives yaks a very different appearance from lowland cattle. The differences extend to their internal organs as well. Their lungs are enormous—three times as heavy as the lungs of the cattle that live in nearby Inner Mongolia, which is a far greater difference than that seen between the lungs of human Tibetans and lowlanders. And, like human Tibetans, yaks have surprisingly little hemoglobin in their blood—less than in lowland cattle.

In spite of these and many other differences, however, one can think of yaks, without much loss of precision, as simply cattle with lots of hair. Zoologists have put them into a separate species (currently most authorities call them *Bos grunniens*), but their gene pool is not clearly separated from that of cattle. Because purebred yaks tend to be ill-tempered and balky beasts, for centuries the Tibetans have hybridized them with lowland cattle in order to produce animals that are a little more manageable. These hybrids turn out to be perfectly healthy, although there have been reports that they are not fully fertile.

All this hybridization has, however, made the task of disentangling the relationship between yaks and domestic cattle difficult. The best way to determine just how genetically separate they are is to compare the DNA of the two species. But it is difficult to find genetically pure yaks. Many animals exhibited as yaks in zoos are actually mixtures. Rather than venturing into the fastnesses of Tibet, scientists have understandably tended to obtain blood from these zoo animals, so they may have been misled into concluding that yaks and cattle are more closely related than they really are.

Still, the general position of yaks on the family tree of cattle and their relatives is undisputed: all the molecular studies show that yaks are by far the closest relatives of domestic cattle that have yet been found. They are much closer to cattle than the Indian gaur and the Javanese banteng, both of which have been suggested as ancestors of domestic cattle breeds. But exactly how much closer they are will remain problematical until this matter of genetic admixture has been straightened out. We may never know, for after centuries or millennia of forced interbreeding with cattle and escape of hybrids into the wild, it is possible that truly pure wild yaks no longer exist.

Further, artificial selection may have been imposed on top of their natural adaptation. The original wild yaks, grazing peacefully on their high-altitude grasslands, probably had no need for such giant lungs and extreme physiology. But when people began capturing them and forcing them to carry immense burdens over the high passes and pull heavy primitive plows through stony soil in the thin air, the yaks best able to carry out these grueling tasks must have been selected.

Selection, whether natural or artificial, is much more likely to proceed rapidly when a population is full of genetic variation. The numerous hybridizations of yaks with domestic cattle could have provided some of this variation, resulting in a kind of evolutionary kick-start.

Humans, may have benefited from the same sort of kick-start. As we have seen, the introduction of genes from lowland human populations into the Tibetan highlands may have helped provide evolutionary fuel for their adaptation to extreme conditions.

There is no doubt that selection is continuing in humans as well as yaks. While infant mortality is slowly decreasing in both Nepal and the Peruvian altiplano (we currently have only fragmentary information from Tibet itself), the rigors of high-altitude life remain severe. The Tibetans, the Quechua, the people of Ladakh, the yaks, and all the other human groups and human-associated species living at the edge of space will continue to evolve better mechanisms to survive these conditions for as long as they continue to inhabit this extreme environment.

It may not be long before some of the genes involved in high-altitude adaptation are tracked down. It recently has been found that one of the two common alleles in the English population for a gene controlling blood pressure is more frequent in people who are able to perform severe physical labor. It is particularly frequent in mountain climbers. Is this allele also frequent in the Tibetan population and less so in the Andean? Or do the Tibetans have yet another allele of this gene that adapts them even better to high-altitude living? The answers will soon be forthcoming.

This is the kind of natural selection that we can easily understand: severe conditions that select for new adaptations absolutely necessary for survival. But what about the rest of us? We are not faced daily with lung-freezing cold or a paucity of oxygen. Still, our lives are not risk-free. Many of us are faced with a different set of dangerous problems that require continuing adaptation. We are only beginning to see just how pervasive selective pressures are, and how difficult or impossible it will be to free ourselves entirely from their rigors. We have not, as many assume, overcome these selective pressures—far from it. Some of them are pressures connected with disease, which are far more pervasive and subtle than we had thought even a few years ago. The next chapter provides a glimpse of some of the difficulties that we still face in our fight for survival against the multitude of organisms that prey on us, and how we have evolved and continue to evolve in response.

FOUR

◈

Besieged by Invisible Armies

*[T]he struggle against disease, and particularly infectious
disease, has been a very influential evolutionary agent,
and . . . some of its results have been rather unlike those
of the struggle against natural forces, hunger, and
predators, or with members of the same species.*

J.B.S. HALDANE, *Disease and Evolution* (1949)

In 1946 my colleague Jon Singer arrived as a young postdoc in the
Caltech laboratory of chemist Linus Pauling. One of his first tasks
was to help in what was then a very tricky problem: to detect a difference between two almost identical proteins.

The project had its genesis in an accidental encounter. A year
earlier Pauling had been invited to join a committee that would eventually recommend, among other things, an enormous infusion of
resources into health sciences after the war. The eventual result was
the emergence of a great research empire, the National Institutes of
Health, that has dominated biomedical research in the postwar years.

In the train on his way to the committee's first meeting in Chicago, Pauling bumped into another committee member, geneticist
William Castle. In the course of their conversation he learned about
a strange disease, found among African-Americans, called sickle cell
anemia. Castle's description of the symptoms of the disease immediately fascinated Pauling.

Although reports of what was probably sickle cell anemia have been traced to as long ago as 1870, the first clear evidence for the disease was found in 1910. In that year James Herrick, a Chicago doctor, noticed an odd phenomenon in the blood of Walter Noel, a black dentistry student from Grenada. Noel's normally round red blood cells would suddenly distort into a sickled shape while they were being examined under the microscope. Herrick and others soon found that many other people of African ancestry had the same puzzling trait.

On the train Pauling's attention was caught when Castle told him why the sickling took place. If a thin film of blood sits for some time on a microscope slide, trapped under a cover slip, the cells in the blood will use up the oxygen. It is only then that the sickling happens, accompanied by an increase in the polarization of light as it passes through the cells. What particularly intrigued Pauling was that as soon as oxygen is reintroduced, the cells quickly snap back into their normal shape and the polarization disappears.

Pauling realized that this visible phenomenon must be due to an invisible molecular event. Since red blood cells are little more than bags of hemoglobin, what was happening to the cells was almost certainly due to some property of the homogeneous population of protein molecules that they contained.

The most likely explanation seemed to be that, in the absence of oxygen, the hemoglobin molecules became linked together in a reversible fashion. These linked protein molecules would polarize light that passed through their fiberlike chains and would actually distort the shape of the cell itself. Then, when oxygen was reintroduced, the chains would break up and allow the normal shape of the cell to reassert itself. Pauling was familiar with such reversible reactions from laboratory experiments. If his idea was right, then the sickling must be a clear signature of a change at the molecular level.

He realized further that people with sickle cell anemia should be carriers of a mutant gene, specifying a mutant form of the hemoglobin protein. Since the dogma current at the time was that genes themselves consist of proteins—only a few heterodox researchers were beginning to turn their attention to DNA—Pauling had no clear idea of the nature of the mutation itself. But he thought that it might be possible to approach the problem by seeing if the hemoglobin protein in sickle cell patients really was different from the normal protein.

Another arrival in Pauling's lab just after the war was a brand-new M.D. named Harvey Itano. In common with most other Japanese-Americans, Itano had begun the war years branded as an enemy alien. He was actually being sent with his family to an internment camp when he received a letter telling him he had been accepted at St. Louis University Medical School. A few months later, through the efforts of many friends, he was permitted to leave the camp and go to St. Louis. But his love was science, and he had no interest in practicing medicine after getting his degree.

As soon as Itano arrived at Pauling's lab, Pauling asked him to look at the sickle cell problem, suggested some physical chemistry tricks that Itano might use, and then promptly disappeared to take up a temporary fellowship in England.

Itano tried Pauling's ideas but got nowhere. In talking over his difficulties with Singer, it seemed to him that the overall electrical charge of the molecule might be a better measure than the ones Pauling had suggested. Luckily, Caltech had a massive apparatus, invented and built in Sweden, for carrying out this sort of experiment. Singer remarked to me with much amusement that he quickly became essential to the project, since he was the only one in the lab who was strong enough to lift the enormous apparatus and lower it into the vast buffer-filled holding tank.

After much experimentation, they homed in on just the right pH of buffer that would distinguish clearly between the electrical charges on the normal and mutant proteins. At a pH very close to neutral, normal hemoglobin migrated toward the positive pole, while hemoglobin from patients with sickle cell anemia migrated toward the negative pole.

They went a step further and examined hemoglobin from relatives of sickle cell anemics, who had no sign of the disease but whose blood cells nonetheless sickled when the oxygen tension was reduced to very low levels. The hemoglobin of these relatives turned out to be made up of a mixture of the two kinds of molecule. This meant that these relatives must be genetic heterozygotes, who had inherited an allele for the normal protein from one parent and an allele for the sickle cell protein from the other. People with the more severe sickle cell disease were homozygotes, who by ill luck had inherited a mutant allele from both parents.

Since the heterozygous relatives did not show the disease, possession of one of these alleles was not enough to produce any medical problems. Possession of two, on the other hand, could be very dangerous. Sickle cell anemia, as Pauling had originally suspected, could truly be called a molecular disease. And it was this memorable phrase that Pauling, on his return from England, scrawled as a title across the top of the paper that Singer had drafted.

A few years later, in 1953, James Watson and Francis Crick announced their famous double-helix model for the structure of DNA. They realized that complementary sequences of bases must be strung along both halves of the helix. Their model explained how genes could be replicated and passed from one generation to the next—each chain of the double helix could serve as a template for a new complementary copy of itself, resulting in two double helices where there had been only one before. But Watson and Crick also realized that each half-helix of the DNA was made up of a long string of bases. Proteins were already known to be long strings of amino acids; both gene and protein could be read like sentences, and the first must code for the second. The language of the DNA was translated directly into the language of the protein. Changes in the sequence of bases along the DNA molecule must be the ultimate cause of changes in the amino acids of proteins.

This reasoning led Crick to suggest to protein chemist Vernon Ingram that he take a careful look at the normal and sickle cell hemoglobin proteins. Crick predicted that the difference between them would turn out to be a small one, and indeed Ingram found that the two molecules were identical in all but one of their 146 amino acids.

Once the genetic code embodied in the DNA itself was worked out, it could be seen precisely how a single change in the DNA could lead to the small change that Ingram had detected in the protein. Tiny as it was, this change could nonetheless have severe effects on people unlucky enough to carry two mutant alleles. All the many billions of copies of the hemoglobin molecule in their bodies would be manufactured with the defect, which would multiply the effect of the mutation enormously.

The discovery by Itano, Singer, and Pauling had huge consequences. Not only did it shed much light on the molecular basis of sickle cell anemia, it helped point the way to an entire new scientific field, that of molecular biology.

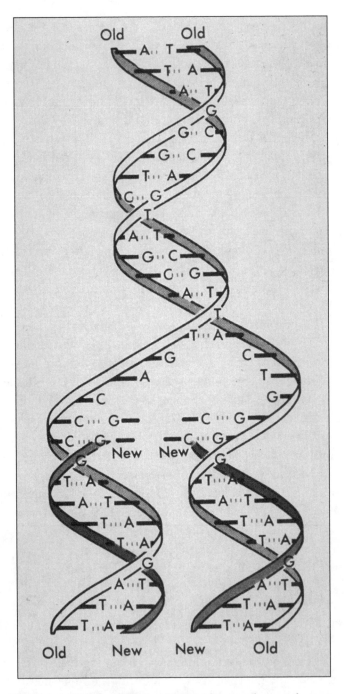

Figure 4–1. The double helix of DNA, showing the complementary bases, the way the code is carried by the molecule, and the synthesis of new complementary strands.

But their finding had still not explained why the mutant gene is so common in the African-American population, and why its effects are so obvious that it had come to Pauling's attention in the first place. We now know the answer to this question: various kinds of complex, rapid, and continuing evolution explain a great deal about the sickle cell and many other puzzling alleles. These evolutionary processes have been accelerated by our cultural inventions, particularly the invention of agriculture. Had we stayed in the forests like our orangutan and chimpanzee relatives, none of them would have happened.

A CLASSIC CASE OF SELECTION

Sickle cell anemia is a serious disease. It is found in one of every four hundred African-American babies in the United States. As the children with the disease grow up, they are susceptible to a series of sudden hemolytic crises, in the course of which large numbers of their red cells are destroyed. These crises can be brought on by something as small as a bout of unusual exertion or even a minor viral illness. They trigger a great variety of symptoms—enormous swelling of the spleen, swelling of the hands and feet, bone necrosis, and sometimes detachment of the retinas. The symptoms tend not to be obvious during the first few months of life, but the cumulative effects of the repeated crises soon build up.

In the 1970s infant mortality from the disease was fifteen percent during the first year of life. Now that the generally white medical establishment has finally realized the seriousness of the disease, this rate has gone down substantially—a 1995 study reported that only twenty of 694 homozygous children had died over a ten-year period, even though there was much serious sickle cell-related illness among the group.

Although some belated progress has been made in palliation of the disease, there has been no dramatic breakthrough in finding a cure. But this is probably only a matter of time. Eventually gene therapy—in which the gene itself is modified or replaced—will hold out hope for a lucky few.

The wide prevalence of the allele for sickle cell anemia is something of a surprise, in view of how harmful the disease is. Of the many different genetic traits that can damage their carriers severely, sickle cell is among the commonest. But the disease itself is only the tip of the iceberg. Heterozygous carriers of sickle cell trait are far more common than people who actually have the disease. Most heterozygotes marry homozygotes for the nonmutant allele, because they outnumber them in the population, and such marriages do not produce children with the disease. Even if two heterozygous people do happen to marry, only one quarter of their children will have the disease. This means that the heterozygous carriers in turn greatly outnumber the homozygotes who have sickle cell anemia. Carriers are in fact forty times as common as mutant homozygotes.

A five percent frequency of the mutant allele—about what is seen in the African-American population—means that about one person in every ten of the population is a heterozygous carrier. Most of the mutant sickle cell alleles in the population lurk almost invisibly in these heterozygotes, and they may be passed on in the heterozygous state for many generations before they come together by chance in an unlucky homozygote.

Even heterozygosity for this mutant allele has its costs. Under extreme conditions possessors of a single mutant allele can suffer a sickle cell crisis. This was first noticed during the Second World War, when African-American soldiers who were being transported in unpressurized planes would sometimes and unexpectedly suffer hemolytic crises at high altitude. More recently, sickle cell trait might have been implicated in a case of sudden death of an African-American athlete. Because this condition is not usually noticed on autopsy, there are likely to have been other cases (although the presence of the allele in one dose does not seem to influence the choice of an athletic career). These harmful effects, uncommon as they are, make it even more puzzling that the mutant alleles are so numerous.

Most of the ancestors of the American children with the disease were originally taken by force from West Africa, where the allele is even more common. In many West African tribes it is found at an astonishing frequency—twenty percent or even higher. Half the members of such tribes are heterozygous carriers of the gene, and five percent of babies are born with the disease—twenty times the rate in the United States. To make matters worse, in some tribes of equatorial Africa there is another allele of the same gene, called hemoglobin C. In combination with the sickle cell allele, it too can also result in serious illness. (This combination occasionally turns up in the United States as well.)

Even with advancing medical knowledge, the devastation caused by sickle cell anemia is still vast. About 150,000 babies are born with the disease every year in sub-Saharan Africa, and even those who have access to adequate medical care face enormous risk from recurrent health crises. Until recently, half the affected children died during their first two years, and very few of them survived to adulthood—though that picture is beginning to change, particularly in the larger cities.

So why, we are driven to ask, in the face of this horrendous illness and mortality, is this damaging allele still found in these populations in such numbers? One would surely expect it to disappear quickly, or at least to fall to very low levels, as a result of natural selection.

In 1949, the same year that Pauling and his group published their paper on the molecular nature of the disease, the eminent geneticist J.B.S. Haldane made a rather hesitant suggestion. Perhaps, he thought, the allele is so prevalent in Africa because the heterozygotes are protected against a blood disease, such as malaria. If this protection were to be sufficiently strong, Haldane pointed out, a kind of genetic balance of death would be set up. People homozygous for the normal allele would be killed by malaria in large numbers, as they presumably always have been. And people homozygous for the sickle cell allele would also die in large numbers—not because of malaria, but because of the effects of the mutant allele that they carry in double dose.

But the story would be different with the heterozygotes. Some of them would die as well—life in West Africa has never been easy. But as long as they did better than the other two groups, then the balance that Haldane suggested would emerge. Even though death will take its toll of all three genotypes every generation, the heterozygotes will be the winners in this grim sweepstakes.

The victory of the heterozygotes will only be temporary, however. Each generation, heterozygous people will meet, marry each other, and have babies. One quarter of these babies will be homozygous for the normal allele and thus at great risk of dying from malaria, and one quarter of them will be homozygous for the terrible sickle cell allele. Only half their children will be heterozygotes and therefore protected. This means that, even though heterozygotes have the highest fitness, they cannot take over the entire population. Even if malaria were so severe that only heterozygotes survived, homozygotes of both types would inevitably appear in the next generation, and the grim toll of death and disease would continue. Figure 4–2 shows how this works.

This kind of balance, known as a balanced polymorphism, has another dramatic genetic consequence. It gives a huge advantage to sickle cell mutations as soon as they first appear. Consider a population that lives in a malarious region, in which all the individuals happen to carry only normal alleles. In such a population malaria will be rampant and highly destructive. Then, if a sickle cell mutation arises and is not lost by chance during the first generation or two, all the copies of the mutation that are passed down to subsequent generations will have an immediate advantage.

The result is that the mutant allele will spread with tremendous speed through the population. It will continue to rise in frequency until the advantage it gives to the heterozygotes is eventually counterbalanced

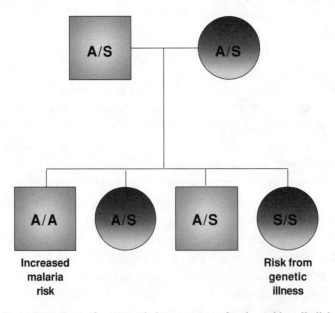

Figure 4–2. Even if two people heterozygous for the sickle cell allele marry and have many children, only half the children will be heterozygous. A quarter will be homozygous normal, and at greater risk from malaria, and a quarter will be homozygous for the mutant allele and will suffer from sickle cell anemia.

by the cumulative disadvantage of the growing numbers of homozygotes. Thus, once a sickle cell mutation arises anywhere in a malarious region, the genetic composition of the population will change swiftly and a balanced polymorphism will emerge.

E. A. Beet, a colonial medical officer working during the mid-1940s in what was then northern Rhodesia, found some evidence for this phenomenon. He noticed that malaria patients who had the sickle cell trait were less heavily infected with the malaria parasite than those who did not. It took decades of further accumulated data to convince the skeptics, but it is now generally agreed that Haldane's inspired guess was indeed right. Heterozygotes for the sickle cell allele really are protected against malaria. But this protection turns out, rather surprisingly, to be a clumsy and ill-constructed shield against the disease.

Contrary to an often-repeated myth, heterozygotes are not protected against getting the disease in the first place. Because the disease is spread by mosquitoes, such protection would imply that carriers of the mutant alleles could somehow fend off the mosquitoes, or could manage to destroy the parasites as soon as the mosquitoes injected them into their bloodstream. They can do neither of these things. Instead, they are protected simply because the severity of the infection is reduced. Even though this protection might seem a case of too little too late, it can often make the difference between life and death.

The gene's effects are exerted most strongly in the bloodstream of the malaria victim. When a malaria parasite enters a red blood cell, the cell is doomed—the parasite's progeny will multiply in the cell, consuming its hemoglobin as a food source. Finally, the cell will burst open and the progeny of the original invading parasite will be released in a jostling crowd, ready to infect other red cells in a kind of biological chain reaction. It is this simultaneous bursting of millions of red cells that produces the fierce fevers of malaria. The icy chills that follow cause the victim to tremble violently, giving the disease its older English name of the ague.

As if this were not enough, the parasites of some types of malaria cause even more damage, starting from the moment they invade the red cells. They change the infected cells' outer membranes, causing them to stick to nearby uninfected cells. As these clumps of red cells accrete, they take on a flowerlike shape known as a rosette.

Rosettes, when they are forced through capillaries by circulatory pressure, can clog them up. The more rosettes, the greater the danger. This helps to explain why the parasite *Plasmodium falciparum*, which produces rosettes, causes the severest kind of malaria. This species multiplies to enormous numbers in the blood, reaching levels several times as high as the other (and somewhat milder) human malarias. The result can be damage to the brain or the kidneys.

The situation is different in people with sickle cell trait. When their red cells are infected by parasites, the oxygen inside the cells is used up as the parasites grow and multiply. In the bloodstream of sickle cell heterozygotes this triggers the sickling reaction, just as it does on a microscope slide. Once the cell is distorted in this way, it is less able to form rosettes with nearby cells. The result is less damage to the capillaries.

An added bonus is that the sickled cell is more likely to be removed from the circulation before the parasites can mature. Our spleen serves as a filter, selectively removing defective red blood cells. The unsickled cells are round, smooth, and flexible, so that they can slip unharmed through the many complex passages inside the spleen. The irregular sickled cells get stuck and are soon destroyed by white blood cells. This filtration has the effect of reducing the total numbers of parasites in the blood, but at the cost of a dangerously swollen spleen.

In spite of all these benefits, there is a point beyond which the sickle cell gene can no longer provide even this crude and clumsy protection. A really severe malaria infection can cause too many red cells to sickle. If the infection is overwhelming, even heterozygous individuals can succumb to the effects of the disease.

THE AGRICULTURAL REVOLUTION AND MALARIA— LIFE AND DEATH GO HAND IN HAND

Sickle cell anemia is such a well-known story in biological circles that, whenever I bring it up in one of my classes, my students' immediate reaction is, "Oh, no! Not this again!" They have already heard about the disease in class after class and assume that by this time there is nothing left to learn about it.

They are wrong, as I gently assure them. The more the story is examined, the more complex it becomes. Our genetic defenses against malaria are influencing the future evolution of our species in an astonishing variety of ways—and perhaps may be affecting the future evolution of the malaria parasites that prey on us as well.

The most important actors in this evolutionary tale are the mosquitoes that spread the disease. People who have done fieldwork in the forests of central Africa find that, provided the forests have not been modified by human activity, their dark and sheltered floors are surprisingly mosquito free. The mosquitoes that are found in the forests tend to live high up in the tree canopy, where they prey on birds, snakes, and arboreal mammals.

The forest mosquitoes themselves belong to a great many different species, each of which is fairly rare, and relatively few of them are inclined to prey on humans. Such diversity among mosquito species is typical of the great variety of animals and plants that are usually found in any undisturbed forest ecosystem, especially in the tropics.

Fields and farms, in contrast, with their plentiful pools of water and disturbed vegetation, swarm with mosquitoes. In equatorial Africa agriculture has had particularly disastrous effects on the environment. The soil in this region tends to be impoverished and filled with iron salts, so that when vegetation is stripped away, it quickly hardens into a red cementlike substance known as laterite. Any ruts and depressions in the ground are made permanent by this crust and tend to fill rapidly with water, providing places for mosquitoes to breed. This problem has been compounded in recent decades by the accumulating detritus of civilization. Abandoned tires, bottles, and plastic containers, filling up with rainwater, provide even more breeding places.

As a result the mosquito population explodes—but not all of the mosquito species found in the region take part in the explosion. Throughout much of equatorial Africa, the new mosquito universe is dominated by a formerly rare species called *Anopheles gambiae*. Currently, this species is responsible for spreading three-quarters of the malaria in Africa. It is particularly common in small villages and outlying farms, wherever rainfall is high. Every inhabitant of such a village may be bitten by malaria-infected mosquitoes several times a night.

This change has been relatively recent and has been particularly marked during the last three or four thousand years. When agriculture, and later the technology of ironworking, were introduced into central Africa by waves of Bantu-speaking peoples who migrated down from the north, the immediate result was widespread deforestation. This increased the incidence of malaria, as the mosquito population altered.

An even more recent and better-documented upsurge in the disease has taken place in the highlands of Madagascar. During the eighteenth century European travelers to this island remarked with surprise on the great contrast between the fever-ridden coasts and the thickly forested, malaria-free highlands. But around the year 1800, following the introduction of rice cultivation to the highlands, this distinction rapidly disappeared. Just as on the mainland of Africa, as soon as the ancient forests began to be destroyed, a formerly rare mosquito species exploded in numbers. This species, *Anopheles funestus*, quickly became the primary carrier of the disease in these formerly healthy parts of the island.

These dramatic, human-caused ecological changes have, like so many human-caused changes, accelerated our own evolution as well. Along with the spread of agriculture and the huge migrations of peoples, the sickle cell allele has risen in frequency. Anthropologist Frank Livingstone has found that the history of the region can be traced in its genetics. For example, tribes speaking the Kwa subfamily of languages have lived in what is now Liberia and the Ivory Coast for a long time, since before the migrations from the north. Near the coast, these tribes have low or even zero frequencies of the sickle cell gene. But the gene is far more common in the eastern regions, where they have come into contact with other invading peoples.

These invasions of the Kwa territory by outsiders have been dated, though with some uncertainty, to the seventeenth century. This means that the gene has had perhaps twenty generations to reach the high levels that are currently found in the contact region. Livingstone suspects that, nearer the coast, the gene has been introduced so recently that it has not yet come into balance between selection for the heterozygotes and selection against the homozygotes.

A ZOO OF NEW ALLELES

The sickle cell mutation is, as we have seen, anything but an unalloyed blessing. It causes a terrible disease in some of its carriers even as it protects other carriers against a different terrible disease. This seems like a very poor bargain. Why has evolution not replaced it with some other allele that confers protection against malaria without entailing such a severe cost?

The sickle cell mutation is only a single mutational step away from the normal allele, and at present in each generation some one hundred brand-new sickle cell mutations are appearing somewhere in the human population. If such a single-step mutation, readily arising somewhere in the population, confers some limited amount of protection, it will immediately be selected for. More "sophisticated" mutations that confer better protection without such damaging side effects are likely to require more than one mutational step, perhaps occurring in the same gene or perhaps in different genes. These multiple mutations need more time to appear, since the probability that they will turn up at all is much lower. Only later, perhaps thousands of years later, will such far less probable combinations of mutations arise. Thus, the apparently clumsy evolutionary response of sickle cell anemia is a function simply of probabilities, not of some evolutionary ineptitude on our part. Given enough time, evolution should be able to solve the problem.

The sophisticated mutations have not yet appeared. Instead, a wide variety of other clumsy mutations have appeared at a variety of genes, some just as harmful as the sickle cell gene. They have been driven up in frequency, in Africa and elsewhere, by the relentless selective pressure of malaria. This zoo of mutations includes other amino acid changes that affect the same hemoglobin gene. They are called hemoglobin C, hemoglobin E, and various other designations. Some of these mutations were originally found by Harvey Itano, who has devoted his life to understanding the many things that can happen to this hemoglobin gene. In many cases we still do not understand what their protective effect might be.

A rather different kind of mutation, widespread in Mediterranean countries, turns the hemoglobin gene off completely. Homozygotes for this gene suffer from a very severe anemia known as thalassemia (from the Greek *thalassa*, "sea," since it is so common among the people who live on the malarious Mediterranean coasts). The people of Greece, southern Italy, and especially coastal Sardinia have high frequencies of this gene.

The parallel with the sickle cell story is striking. Thalassemia, too, is held in the population as a balanced polymorphism, and the protection against malaria that it confers on heterozygotes is equally clumsy. Turning the gene off does not make all the hemoglobin in the red blood cells disappear, since other hemoglobin genes are still active. But it does damage the hemoglobin molecule itself, which tends to shorten the lives of infected red cells so that they are destroyed early. Once again the protection comes at the cost of severe anemia and swollen spleens.

In Chapter 2 we met the G6PD gene, which has dangerous effects if its carriers eat broad beans. Hundreds of different mutations of this gene, which codes for an enzyme that is very important in red blood cells, have been found throughout the Mediterranean and parts of Africa. The mutant alleles are particularly common among the little villages of Greece. Sometimes the inhabitants of villages only a few kilometers apart can carry quite different mutations. Again, the damaged genes shorten the lives of red blood cells and diminish the effects of malarial infection.

In New Guinea and Micronesia a completely different kind of mutation has appeared that also protects against malaria. It changes the properties of the membrane that surrounds the red cell, altering its shape from a circle to an ellipse. This change makes the cell more resistant to the entry of the malaria parasite. Because the mutation actually increases the resistance to the parasite itself, it would appear at first sight to confer protection that is less clumsy and damaging than sickle cell or thalassemia. Unfortunately, these elliptical cells are not as flexible as normal cells, so the spleen detects them as abnormal. Homozygotes for this gene can also suffer from anemia.

This list is not exhaustive by any means. It is striking that all these mutations appear to be half-baked and jury-rigged. They are exactly the kind of "easy" mutations that one would expect to be selected for immediately in a population that has been subject to a brand-new and severe selective pressure. And indeed, there is genetic evidence that many of this zoo of new alleles have arisen in the last few thousand years—which is just the period during which we human beings have had a substantial impact on our environment.

MILD OLD DISEASES AND SAVAGE NEW ONES

In a perverse way these various protective mutations may actually be making things worse for us, at least in the short term. In a kind of evolutionary warfare, every genetic move that we make is countered by a move made by the parasites. Although this situation is temporary, it has contributed substantially to our rapid evolution.

The received wisdom (to which, by and large, I subscribe) is that long-continued host-parasite relationships will eventually lead to milder forms of a disease, because hosts and their parasites will co-evolve. Strains of parasites less damaging to their hosts will be selected for, since this lengthens the period of time during which the parasites can be passed on to new hosts before the old hosts die. Strains of hosts more resistant to their parasites will also be selected for, which will make it more difficult for the parasites to spread. When this happens, fewer parasites can survive in any individual host. But although the chance that the parasites can spread to new hosts during any given time period may be decreased, the very mildness of the disease lengthens the window of opportunity during which the parasites can spread.

Epidemiologist Paul Ewald of Amherst College argues that this pattern does not always hold true. Under some conditions highly pathogenic strains of disease organisms can survive or even thrive, so that these strains will continue to be selected for. Suppose a pathogen spreads easily: Then, even if it kills its host quickly, it can go on spreading so long as there is a plentiful supply of other hosts nearby.

I agree that this can happen—the history of our battle with diseases is rife with examples of selection for high virulence. Perhaps the most extreme of these was the terrible Black Death. Many other diseases have swept through our species during historical times with disastrous results. But this apparent plethora of extremely virulent diseases is an artifact, the result of our understandable tendency to concentrate our attention on our own diseases. When we look at the natural world at large, highly virulent diseases are not so common. Animal and plant plagues, while not unknown, appear to be rare in undisturbed ecosystems.

The reason the situation is very different for us is that we are the planet's preeminent disturbers of ecosystems. An ecological disturbance, such as the introduction of agriculture to central Africa, can lead to the temporary success of a poorly adapted pathogen. The pathogen can survive even if it causes a severe disease in its human hosts, because the hosts are now so numerous. As a result of our efforts to change the world, we (and our domesticated animals and plants) have suffered dreadfully from diseases.

Our rapid cultural change has driven not only our own evolution but the evolution of our parasites as well. This becomes vividly apparent when we compare the malarias that we suffer from with the malarias that afflict our far more slowly evolving relatives, the chimpanzees.

Like humans, chimpanzees have at least four different malarias, but they cause much milder forms of the disease. Zookeepers who have examined chimpanzees freshly captured from the wild have often found malaria parasites in their bloodstreams, but the chimps do not seem incommoded by their presence.

The nearest known relative of our own *falciparum* malaria (and not a very close one at that) is a chimpanzee malaria parasite called *Plasmodium reichenowi*. DNA evidence suggests that that the *falciparum* and *reichenowi* plasmodia have each had long histories of association with their unwilling hosts.

If malaria in humans is such an old disease, and chimpanzee malaria is so mild in its effects, then the conventional wisdom would predict that *falciparum* should also be mild in us. Instead, it is vicious. A likely explanation for this contradiction is that our disease has not always been so nasty.

Before the arrival of agriculture, *falciparum* parasites must have already infected the peoples living in the central African forests. We do not know how serious a disease they caused, and we do not know which species of mosquitoes were responsible for spreading the parasites through those sparsely populated and heavily forested regions. But it seems likely that they were not the same mosquitoes as the ones that spread the disease today.

Perhaps the unknown mosquito vectors that once carried the parasites from human to human in the forest have disappeared or become confined to the few remaining undisturbed regions. In order to survive this sudden change, the parasites had to adapt to one or more of the relatively small number of mosquito species that are able to thrive in open and disturbed country. It would be highly unlikely that these mosquitoes would be particularly good at transmitting the parasites.

Indeed, from the parasite's point of view, today's mosquitoes do a rather mediocre job. *Falciparum* malaria is transmitted poorly by its most numerous host, *Anopheles gambiae*. Only a small fraction of the mosquitoes of this species that drink *falciparum*-infested blood manage to transmit the parasites to new human hosts. Further, as was recently shown by Jakob Koella of the University of Åarhus in Denmark, infected mosquitoes live a shorter time than uninfected ones. Even though infection does stimulate the mosquitoes to bite more victims, they are on the whole poorly adapted to the parasites.

These days, therefore, the parasites must multiply in swarms in the blood of their victims. Otherwise, their inept mosquito vectors would not be able to pick them up in numbers sufficient to ensure that some of them will be passed on to other human hosts.

The risk to the parasites of this uncontrolled multiplication can be high, for they might kill their current hosts from cerebral malaria or other severe symptoms before they manage to spread further. But the benefit also seems to be substantial, since their increased virulence allows them to spread through the crowded host populations in which they now find themselves.

We have no direct proof of a recent increase in the level of virulence in the *falciparum* parasite. Nor have we any firm information about what *falciparum* malaria might have been like before the advent of agriculture. But it is striking that *reichenowi*, the chimpanzee relative of *falciparum*, is so mild by comparison. Moreover, additional genetic evidence shows that chimpanzees, unlike humans, have not been forced by ecological upheaval to fight an escalating evolutionary war against their parasites. No equivalents of the human sickle cell and thalassemia genes have been found in chimpanzee populations; such clumsy, half-baked protections have simply not arisen there.

The transmission process, too, is likely to be very different in chimpanzees. Nothing is currently known about the mosquitoes that transmit *Plasmodium reichenowi* from one chimpanzee to another. But when these mosquitoes are found and tested, I confidently predict that they will turn out to be excellent carriers of the parasite. Even though they must be rarer than the clouds of *Anopheles gambiae* that swarm in the fields and villages where most humans now live, they still manage to spread *reichenowi* among small and scattered bands of chimpanzees. The *reichenowi* parasite must be superbly adapted, not only to chimpanzees but also to the mosquitoes that carry it readily from one animal to another.

We should not be surprised, therefore, that the disease seems to have so little effect on the chimpanzees that acquire it. The chimpanzees and their malarias have reached an evolutionary compromise, in which ready transmission has been accompanied by lowered virulence.

Every year brings a new discovery about how clever chimpanzees are, but they are not yet clever enough to invent agriculture. They have lived in forested regions in central Africa for a very long time, perhaps for several million years. During all that time, both they and their environment have remained relatively unchanged. They have thus had plenty of time to reach such accommodations with their parasites. (They have lots of other parasites, by the way, in addition to those that cause malaria.) But recent human-caused ecological disturbances are already taking their toll. As Jane Goodall observed at her site at Gombe Reserve in East Africa, chimpanzees have proved to be tragically susceptible to several newly introduced human diseases, notably polio. Through no fault of their own, like so many other endangered species, chimpanzees are succumbing to human alterations in their environment.

KEEPING UP WITH THE PATHOGENS

While the sickle cell allele and other mutations provide limited protection to the human populations that carry them, they make it more difficult for the malaria parasites to spread. In turn, this drives the parasites to make an evolutionary response.

Some West African populations, as we have seen, have a twenty-five percent frequency of the sickle cell gene, and almost half the people are heterozygous carriers. The parasites that happen to invade heterozygous carriers are immediately put at a disadvantage, compared with the luckier parasites whose hosts do not carry the sickle cell allele.

The disadvantaged parasites will be able to survive and spread only if they have mutated to higher-than-average virulence, allowing them to overwhelm their victims' defenses. This increased virulence will raise the genetic stakes, selecting for even more half-baked, partially protective mutant genes like sickle cell in the host population. And this will in turn select for even more virulent strains of parasite. Thus the spread of agriculture is not the only reason for increased parasite virulence: the genetic response of the human population has contributed to the problem as well.

Recent work has shown that the alleles are doing even more damage to the human population. *Anopheles* mosquitoes that have fed on the blood of malaria-infected sickle cell carriers are more likely to spread the disease than those that have fed on malaria-infected blood that has normal hemoglobin. In sickle cell heterozygotes the parasites must multiply to huge numbers if they are to be picked up by the mosquitoes, but if they are picked up they are more likely to spread. This is another way in which the sickle cell allele may do more damage than it prevents.

Leigh Van Valen of the University of Chicago has described such evolutionary races in terms of Lewis Carroll's Red Queen. In *Through the Looking Glass* the Red Queen remarks that in her land (where everything is reversed) everybody has to run as quickly as possible just to stay in the same place. Van Valen points out that very often in the past we have had to run furiously in an evolutionary sense simply to keep up with our pathogens.

If we invent some technological means to conquer malaria, the terms of this race will change completely. But suppose that we do not: How might we and our parasites break free of this vicious tit-for-tat? One obvious escape route is for the parasites to become better adapted to their mosquito hosts. Doing so would relieve some of the pressure, halting the mindless escalation in parasite numbers. This already seems to have happened with the other human malarias, which have been shown to survive better in their mosquito vectors than the *falciparum* parasite can manage to do.

Another escape route might open up if a new mutation appears in the human population that confers strong protection against *falciparum* malaria without the terrible cost of sickle cell. Just such a mutation, which bestows essentially complete protection against a less dangerous form of malaria called *vivax*, has already appeared in central Africa. People homozygous for this mutation, called Duffy-negative, lack a protein on the surface of their red cells that aids the entry of the *vivax* parasite. In contrast to sickle cell homozygotes, the homozygotes for Duffy-negative seem to suffer no ill effects—people can, it seems, get along quite nicely without this particular protein.*

Since homozygotes for Duffy-negative are not obviously harmed, no balanced polymorphism slows the spread of the allele, with the result that it has spread through much of central Africa. The incidence

---

* What does the Duffy-negative protein do when the *vivax* parasite is not binding to it? Nobody is quite sure, but we do know that it resembles proteins that are involved in the inflammation response.

of *vivax* malaria has been driven down in these regions, but this does not mean that the Red Queen race between ourselves and the *vivax* parasite has been halted. It is only a matter of time before a mutant arises in the parasite population that will allow the parasites to enter the blood cells by a new route, which will circumvent the Duffy-negative protection and start the whole grim cycle all over again.

Over the last few thousand years, the invention of agriculture has dramatically accelerated our evolution. But the Red Queen race against malaria that has formed such an important part of this acceleration may be coming to an end. When a good immunization against malaria is developed—as it surely will be soon—then the many genetic changes in our species that currently provide clumsy and dangerous half-protections against the disease may all be rendered moot.

Before we assume that the conquest of malaria and other diseases will halt all such evolution, however, we must remember that these dangerous diseases are sure to be replaced by others.

THUNDERING HERDS OF RED QUEENS

Malaria is not the only story of this kind. As a species, we find ourselves in the middle of dozens or perhaps hundreds of such evolutionary Red Queen races that involve many different diseases. Let me give just a few examples.

AIDS has spread from Africa to many other parts of the world in a mere two decades. Although it appears to be a very new disease in our species, having recently made the jump from other African primates to us, evidence is growing that it is not unique. It seems that we have been caught up for a long time in a Red Queen race with similar viruses—a race that is just as complex as the race against the malaria parasites.

One reason that AIDS infections cannot currently be cured is that the viruses are able to incorporate themselves directly into our chromosomes and lurk there for years. The RNA of the AIDS virus is converted to DNA, then slips in among our own genes. This process reverses the normal flow of genetic information in the cell, which goes from DNA to RNA. It is because of this property that such viruses are called retroviruses.

Although this ability makes the AIDS virus almost impossible to eradicate from an infected host, it also provides us with a clue to the prevalence of retroviruses in our own past. Thousands of bits of "fossil" virus DNA lie in our chromosomes, consisting of the much-damaged remnants of old retroviruses. They constitute strong evidence that we have met retroviruses very much like the AIDS virus in the past.

Moreover, we probably waged a Red Queen race with them. Stephen O'Brien of the National Cancer Institute has recently found that people of European ancestry carry an allele conferring strong protection against the commonest AIDS virus. This protection works in a way that is very similar to the protection that the Duffy-negative genotype provides against *vivax* malaria. People who are homozygous for the resistance gene lack a protein that facilitates the entry of the virus into cells.

About one percent of the population is homozygous for this protective allele, and almost all of this one percent are completely resistant to the virus. Far more people—about twenty percent—are heterozygous; they are also somewhat protected against the consequences of infection. Unlike homozygotes, heterozygotes can be invaded by the virus, but the invasion seems to be more difficult. Thus, although infected heterozygous people will eventually develop full-blown AIDS and succumb to an unstoppable tide of opportunistic infections, the course of the disease will be significantly slowed.

O'Brien speculates that an earlier retrovirus, now vanished, drove the resistance allele up to such high frequencies in Europe. Or perhaps the selective pressure that drove the allele up was due to a completely different disease that just happened to use the same entry mechanism into cells. The disease might have vanished, or it might still be among us. Whatever the disease was or is, it seems to have been most active in Europe, since the resistance allele is found there much more commonly than in the Middle East and India.

O'Brien suspects that the earlier disease was probably different from AIDS in one respect. The protective allele (and a protective allele of another gene that has been discovered more recently) are common among Europeans but appear to be rare in Africa. The European disease, whatever it was, probably did not originate in Africa. If it had, then the protective alleles would have been selected for there as well. It appears to be a coincidence that the same alleles confer protection against the new threat posed by the AIDS virus.

HIV resistance is also seen in Africa, but its genetics is still mysterious. There, too, a minority of people have been found who cannot be infected by the AIDS virus, but the gene or genes responsible have not yet been tracked down.

Does this mean that the AIDS virus has been loose among humans in Africa for longer than we thought? Or that other viruses or bacteria, perhaps similar to those in Europe, have been involved in selection for these unknown alleles? Even in Europe the resistance alleles that have been found so far account for only a minority of the people who are resistant to the AIDS virus, which means that many other genes must be involved. If O'Brien is right, then AIDS is only one of a large set of similar diseases that we have battled against in the past. And very recently, he and a large group of colleagues have traced one of the HIV-resistance alleles to an origination time of roughly 700 years ago. Could the Black Death have been the selective agent that drove up the allele in frequency? Experiments are underway to test this possibility.

AIDS-like viruses and malaria are just some of the diseases to which our species has responded genetically. We seem capable of carrying on such genetic wars on many fronts simultaneously. Other fascinating relationships between genes and diseases are coming to light, particularly now that it is possible to carry out experiments that were undreamed of just a few years ago. Consider the debilitating, and ultimately fatal, disease known as cystic fibrosis (CF for short).

CF results from a fundamental difficulty with chloride transport across cell membranes, which triggers a variety of symptoms. The most serious, and the one from which the disease gets its name, is the formation of destructive cysts in the pancreas. In the days before CF patients could receive the missing pancreatic digestive enzymes, these cysts were inevitably fatal.

If enzyme supplements have lengthened the life spans of victims of the disease, they have also provided time for new and more insidious symptoms to appear. The worst of these is gummy mucus that forms in the lungs, gradually blocking off air passages and allowing bacteria to multiply. Eventually, and inevitably, these repeated bacterial infections destroy the lungs.

About one European baby in 2,500 is born with CF, which means that about one person in twenty in the European population is a heterozygous carrier of the mutation. CF is so common among

Europeans that it is sometimes called the Caucasian sickle cell anemia, though there seems to be no connection between the disease and malaria.

CF can be caused by many different mutational changes in the gene. One such altered allele accounts for about a third of the cases, and seems to have been in the European population for at least ten thousand years. Perhaps some kind of heterozygote advantage kept it and other CF alleles there. But against what diseases do CF alleles protect?

The suspected villains are not exotic tropical infections but rather are common diarrheas, the chief cause of mortality among children and the elderly in less developed parts of the world. The worst such diarrhea, without doubt, is the terrible scourge of cholera, which has caused worldwide pandemics since the eighteenth century and still breaks out periodically.

These diarrheal diseases result from a sudden rush of body fluids into the intestine. Bacterial or viral toxins trigger the rush by changing the electrolyte balance of nerve cells in the walls of the intestine which control the fluid transfer. It has been suggested that nerve cells in the intestines of CF heterozygotes, with their defect in chloride transport, are less sensitive to the triggering signal.

It would hardly be possible to infect carriers of the CF gene with a disease such as cholera for research purposes. Luckily, molecular biologists have recently provided some remarkable and suggestive evidence by using mice.

A mouse model for CF has now been made by destroying the mouse equivalent of the human CF gene and replacing it with normal or mutant human genes. Mice that are homozygous and heterozygous for the mutant CF allele can now be bred in any numbers in the laboratory. The homozygous mice are not a perfect model, since their lives are too brief for them to develop more than the earliest stages of the disease. The heterozygous mice, like heterozygous humans, show no obvious symptoms. But these mutant strains can be used in many experiments.

Sherif Gabriel and his colleagues at the University of North Carolina examined the sensitivity of these mouse strains to the toxin produced by the cholera bacillus. Mice that did not carry any human CF alleles quickly succumbed to the toxin's effects, but mice that were homozygous for the human CF mutation were completely resistant—their intestines functioned normally. Heterozygotes, although affected by the toxin, did not become as ill as normal mice. Just as with heterozygotes for the sickle cell and AIDS-resistance alleles, the heterozygous CF mice were not protected against getting the disease in the first place but seemed protected against its severest effects.

Cholera toxin was the obvious choice for these initial experiments, because its effects are so powerful. But the cholera bacillus, which only appeared in Europe in the early nineteenth century, is only one of a large army of bacteria and viruses that cause diarrhea around the world. In the bad old days in Europe before there were any notions of public health, such diseases must have been even more powerful selective agents than they are today. Very recently, researchers in England and the United States have shown that the typhoid bacillus attaches to the CF protein as it enters intestinal cells—typhoid is therefore probably one of the diarrheal diseases that helped to select for the CF allele in Europe.

Problems remain with the theory connecting CF with diarrhea. For example, why are CF mutant alleles not common in the tropics, where diarrhea is even more widespread than it is in the temperate zones? One would expect the alleles to be at particularly high frequencies in Bengal, where most of the great cholera pandemics of the last three centuries have originated. There, cholera has devastated the population for centuries and even millennia—providing a very long span of time for selection to have driven up the CF allele frequencies. Yet CF mutant genes do not seem particularly common in that part of India.

Paul Quinton, a physiologist at the University of California at Riverside, has an ingenious explanation for this discrepancy. One striking symptom of CF is a high concentration of salt in the sweat. Indeed, a "sweat test" for salt is commonly used to detect the onset of the disease in children. Quinton suggests that in hot countries heterozygotes for CF might have been selected against because they would lose dangerous amounts of salt through their sweat. This theory makes a lot of sense—we take salt for granted now, but for most of human history and for most groups of people, it has been in desperately short supply. Such selection, since it would operate each hot season, would outweigh any advantage that CF's resistance to diarrheal diseases might confer. CF heterozygotes who happened to live in a cold region like Europe would not suffer so much salt loss, and their resistance to diarrheal diseases could then become important.

Quinton's theory is ingenious, but there are still difficulties. For example, it is not obvious why CF mutations are very rare among the Chinese, many of whom live in a relatively cool climate. In spite of these puzzles, however, evidence is growing that CF, like sickle cell and thalassemia, is held at high frequency because it forms a balanced polymorphism.

ECOLOGICAL AND GENETIC UPHEAVAL

As our environment changes, the situations that have led to these various balanced polymorphisms are likely to disappear. In the next century we will likely stamp out many of the most contagious and frightening human diseases, just as we have already conquered or nearly conquered smallpox, chicken pox, and polio. But four great factors remain that will lead to further disease-driven evolution.

The first and perhaps most important factor is the continuing global ecological upheaval that we are causing. In the next hundred years most of the natural world will have either disappeared entirely or become drastically simplified. In the introduction to this book we glimpsed some of these changes taking place in Borneo. Ecological changes in central Africa have resulted in the emergence of AIDS viruses. During our lifetimes and those of our children, more such epidemiological emergencies are likely to come thick and fast. Newly resurgent dengue hemorrhagic fever and yellow fever are recent examples.

Second, most of the disease-caused mortality in the world today is the result of common respiratory and diarrheal illnesses that chiefly affect the very young and the very old. These diseases are most prevalent in the developing world, and it will take a very long time before improved sanitation and nutrition finally conquer them.

Third, our technology is also producing many environmentally driven Red Queen races. The indiscriminate use of antibiotics and antimalarials is producing new generations of resistant pathogens. Some people will be better able to survive these resistant strains than others, and our gene pool will change as a result.

Fourth, even in the developed world pathogens play a larger role in disease than we had imagined until recently. Protozoa, bacteria, and viruses that were formerly unknown, or were formerly supposed to be benign simply because they are found everywhere, are now suspected to make a contribution to many different illnesses that were previously attributed to poor diet, unwise living habits, or psychological factors. Even though most of these pathogens may not cause obvious and dramatic symptoms, they nonetheless affect our health and reproductive abilities.

One recent example that has gained worldwide attention is the bacterium *Helicobacter pylori*, which is now known to be responsible for many stomach and duodenal ulcers. Ulcers may not at first glance seem life-threatening, but in the underdeveloped world they often provide a route for systemic bacterial infection. Throughout much of the world *Helicobacter* is now being revealed as a very dangerous organism indeed.

It is also now being discovered that even heart attacks are influenced by bacterial and viral infections. Although heart disease is of course correlated with such factors as fatty diet and lack of exercise, these are not the whole story. There are persistent and growing reports that the formation of arterial plaques may be triggered by the primitive but widespread bacterium *Chlamydia pneumoniae* and by the equally ubiquitous cytomegalovirus. In fact, the rise in heart attacks since the beginning of this century may not even be due to our current terrible diet after all—diets in the eighteenth and nineteenth centuries tended to be even fattier and less balanced! Some of the rise may be the result of crowding and of ecological changes that have allowed these plaque-triggering organisms to become more widespread in our population.

Very recently a group of Swiss researchers discovered that if they infected mice with a specific mouse retrovirus, the infection could trigger the subsequent destruction of pancreatic islet cells, producing diabetic symptoms. In humans this type of diabetes has many of the properties of a so-called autoimmune disease, in which we make destructive antibodies against our own pancreatic cells. But the trigger that first stimulates our immune system to produce these antibodies remains elusive. It now seems possible that the trigger in humans will also turn out to be a viral infection. The reason the connection between infection and diabetes was not found earlier might be that the virus tends to come and go early in life, long before damage to the pancreas begins to develop.

A variety of puzzling illnesses may also be associated with infectious agents, including autoimmune diseases like multiple sclerosis and perhaps even such hard-to-define conditions as chronic fatigue syndrome (CFS). The jury is still out on whether this syndrome is real or entirely psychosomatic, but many people with CFS are convinced that their suffering has a physical cause. Some studies have suggested there may be as many as two hundred cases of the syndrome for every hundred thousand people, though it is difficult to obtain reliable information about such an amorphous illness. Complicating things further, it has been found that simply telling a patient that he or she might have CFS causes the patient to feel worse. The act of putting a name to the disease can make it seem more real. Several viruses have been implicated in CFS, ranging from bornaviruses through various herpes viruses to cytomegalovirus, but the smoking gun has yet to be found. Indeed, the presence of a virus alone does not mean that it causes the syndrome.

Such a syndrome, so difficult to diagnose and variable in its effects, might have an almost invisible selective effect. Let us suppose that at least some cases of CFS turn out to be due to the spread of pathogenic agents. As the disease drains away energy and initiative, does it make infected people less likely to reproduce? If it does, then such selection could be powerful and yet at the same time virtually undetectable.

Other plagues, not necessarily associated with disease organisms, are becoming more important in our lives—ironically, just as the role of extremely severe infectious diseases would appear to be diminishing in the developed world.

Addictions to tobacco, alcohol, and drugs are obvious examples. It can be argued that mortality from tobacco happens too late in life to have much impact on fitness in an evolutionary sense. But like CFS, tobacco-related illnesses can affect people at every age and may also affect their reproductive abilities. The high rate of mortality and severe injury among young people that result from some of these addictions must also be having an effect, not just on the victims but on the gene pool of our species. The related plague of violence, which in the United States and in many developing countries has been carried to an extreme, must also be having an impact. It too is a disease, though its causes are more psychological than physical. Death by gunshot or a drug overdose can be just as effective selective agents as death by infectious disease.

As we conquer some of our obvious disease enemies, the mix of those to which we are still subjected will change. The effects of disease-caused selection may become less obvious with time, but they are sure to continue. And the frequencies of many different alleles in our gene pools will shift in compensation, as we run these many subtle but still threatening Red Queen races.

When we first think about diseases and evolution, it seems obvious that the selective pressures involved ought to involve life-or-death situations, nature red in tooth and claw. But the ramifications of even such dramatic types of selection can be very widespread and subtle. Further, the terms of many of these evolutionary battles have recently changed. As we have "subdued" the natural world, some of our pathogens have become even more dangerous. There is no escaping from the consequences of such natural selection—Darwin is, even from beyond the grave, continuing to change our lives.

FIVE

◈

Perils of the Civil Service

*The perfect bureaucrat everywhere is the man who
manages to make no decisions and escape all
responsibility.*

BROOKS ATKINSON, *Once Around the Sun* (1951)

Sometimes selective pressures can arise as a result of our daily interactions with other members of our society. As our social environment
increases in complexity, these selective pressures will become stron-
ger. In this chapter and the next, we will turn from the physical sources
of selective pressures to the far more subtle psychological sources.

The effects of psychological stress on health are sometimes profound, but few studies have examined the long-term consequences of
such selection on our species. In some cases, however, we can get a
glimpse of the selective pressures involved.

The University College School of Public Health stands on a quiet
London side street, not far from the great complex of university and
medical research buildings centered on Gower Street. Here, in a special
suite of rooms, epidemiologist Michael Marmot and his colleagues have
been examining some of these subtle psychological interactions, in particular how position in the social hierarchy can affect a person's health.

At the nearby School of Tropical Medicine and Hygiene, doctors and epidemiologists wrestle daily with the problems of malaria
and schistosomiasis in far-flung parts of the world. Marmot's work
deals with far less threatening but nonetheless real situations that are

much closer to home. His group has, over a span of three decades, carried out two massive studies of health in the British civil service. During a recent visit I talked with his group about the evolutionary implications of the studies, which have become known as Whitehall I and II.

Whitehall is the ancient street that runs between Charing Cross and the Cenotaph; it passes through the very center of English government. The whole area is filled with government buildings, each with its own maze of offices. Here tourists who have strayed off the beaten track find themselves in a world of bowler-hatted bureaucrats, clad in discreetly tailored suits and dark overcoats and carrying tightly furled umbrellas. Sometimes top functionaries can be glimpsed as they are whisked from one important meeting to another in black Bentleys and Jaguars.

Surely, one would think, this is an unlikely place to detect the ravages of natural selection. Nonetheless, Marmot and his colleagues may have found distinct traces of it at work among these hurrying civil servants.

One of the many cultural differences between England and the United States is that Americans tend to worry about societal inequalities that have resulted from racial prejudice, while the English worry about those that have been generated by the class system. These two causes of inequality are entangled in both societies, but sometimes a story emerges in which the effects of one or the other stand out in vivid relief. The Whitehall studies are particularly striking examples.

THE MINISTRY OF FEAR

The British civil service, like that of many other countries, is a hierarchical system in which pay and responsibility are carefully graded. At the top are the permanent secretaries and undersecretaries, who serve under an ever-shifting collection of politically appointed cabinet ministers and provide the government's continuity (and inertia—you may remember Sir Humphrey, the permanent secretary in the wonderful television series *Yes, Minister,* who used every art at his command to ensure that nothing ever got done).

In the middle regions of the hierarchy are vast numbers of functionaries, along with their secretaries and clerks, and at the bottom are the people who sort mail, carry messages, and perform other simple tasks.

None of the work is physically demanding, job turnover is very low, and every employee has access to roughly the same level of health care through the National Health Service (though most of those at the top supplement it with their own private doctors or additional health plans). The salary differential is steep, though not as precipitous as at a private company—the permanent secretaries at the top make about twenty times as much as the employees at the bottom.

The first Whitehall study dealt only with males, but the second was expanded to include women. In both studies the civil servants who staffed twenty different government departments were divided into six categories depending on their positions in the hierarchy. They were asked to fill out detailed questionnaires about their health, job satisfaction, financial situation, and degree of familial and social support. In addition they were given electrocardiograms, their blood pressures were measured, and they were subjected to a variety of other medical tests.

During the period 1967–69, data were collected from the first cohort of 17,530 men. Then the cohort was followed for the subsequent ten years, in order to see what would happen to them. Data began to be collected from the second cohort of 6,900 men and 3,414 women almost twenty years later, during 1985–88. The second cohort is still being followed, and so far the patterns that are emerging in the males are remarkably similar to the patterns seen in the first cohort. The inclusion of women, however, has added a whole new dimension. It was this new data that has revealed the possible evolutionary implications of the second Whitehall study.

The most dramatic pattern to emerge from both studies is that, once age is factored out, position in the hierarchy is strongly correlated with the likelihood of mortality. With each step down the hierarchy, the likelihood of death increases. When the bottom and the top categories are compared, the differences are surprisingly large: a civil servant in the bottommost category is three times as likely to die during a given period as one of the same age who happens to occupy a position at the top.

This greater mortality is correlated with a number of obvious health-related factors. In both studies, with each step down in the hierarchy, smoking becomes commoner, as does obesity. Diet tends to be less healthy at the bottom end, and both men and women are less likely to exercise regularly. Signs of early heart disease are more common in the lowest groups.

These factors are not the whole explanation for the increased mortality, however. This is because the physical risk factors tend to be more concentrated at the bottom of the hierarchy, while the risk of mortality increases with each step down.

For example, when the Whitehall II cohort of civil servants was tested during 1985–88, detailed ECG measurements turned up indications of incipient ischemic heart disease—the effects of narrowing of the coronary arteries—in about five to six percent of the people in each of the top five categories. But the signs of ischemia jumped to 10.5 percent among those in the sixth and lowest job category.

In contrast, when the top administrative category is compared with the category of professionals and executives just below it, even this small step down the hierarchy results in a highly significant increase in mortality. People in this slightly lower-ranked group, all of whom are still at the managerial level, are about one and two-thirds times as likely to die (again, once age is factored out) as those in the top group.

Marmot discovered that mortality rates seemed linked less closely to physical risk factors than to the mindset of the study participants. When the members of the Whitehall II group were asked to self-rate their health, the ratings were again correlated linearly with position in the hierarchy, rather than reflecting the health risk factor. The proportion of those rating their own health as average or below rose steadily from 15.3 percent in the highest job category to 33.7 percent in the lowest.

It would seem that, particularly among the highest categories on the scale, the differences in mortality are not due to physical risk factors—which leaves psychological factors. Even an artificial construct like the civil service system can apparently create its own selective effects.

More recently, Marmot and his collaborators carried out two studies that reinforced their conclusion. In the first they concentrated on one of the twenty government departments, singling it out because it had just become widely known that the department was scheduled to be privatized. This traumatic news caused the incidence of self-reported health problems to shoot up immediately, particularly among the men, although during the same period these measures showed no change in the other nineteen departments.

In the second study they examined death rates over the long term. They returned to the first Whitehall cohort, for which statistics had been collected in the 1960s, and examined what had happened to its members after they retired. Of course, mortality kept going up as the cohort got older. But the researchers wanted to find out how well the job category that a civil servant had occupied before retirement predicted the likelihood of death during the study. What happened, in short, once the job-related pressure was off?

They found that hierarchical position was still a predictor, but only about half as good as it had been among the civil servants who had met their demise while still in harness. After the psychosocial stress had eased, the gaps in mortality between the higher and lower job categories tended to close.

Unfortunately, this study was confounded by the probability that the people who had survived to retirement, regardless of their hierarchical level, were those who by definition were less likely to succumb to work- and status-related pressures. Yet the weight of evidence is that the psychological pressures are real and affect some people disproportionately. Another piece of evidence came from cases of coronary heart disease among the Whitehall II cohort, cases that had turned up since the first statistics were collected on the cohort. These new cases, Marmot's group found, were most common among those civil servants who felt they had little or no control over their jobs, regardless of which job category they occupied.

MENTAL HEALTH AND EVOLUTION

The health implications of the Whitehall studies and others like them are far-reaching. They are unusual among sociological studies in that they clearly factor out and demonstrate the importance of psychological influences on physical health.

Earlier studies had shown clearly that in most societies position in the socioeconomic hierarchy has a huge influence on health, but it had proved difficult to disentangle the psychological and physical effects of being low on the socioeconomic scale. The Whitehall studies have been able to remove at least some of the influence of physical factors. They show that, at least in the upper echelons of the Whitehall social hierarchy, position has a greater impact on health and survival than mere income. The far less visible, and thus far less easily measured, psychological selective pressures are created by the society of which the individual is a part and are perceived by individuals in many subtle ways. Some of these ways, it seems, can have a detrimental effect on health.

To be sure, these psychological pressures act in tandem with physical factors, and the effects of the two may often be multiplied together. While it used to be fashionable to attribute heart attacks to diet or to a hard-driven, overachieving "type A" personality, we now know that pathogenic organisms may also play a role. To make the story even more complicated, it is now becoming apparent that a remarkably good predictor of the likelihood of a heart attack is a person's mental health. Twenty to thirty percent of people who have suffered myocardial infarcts, and an astonishing forty percent of people with coronary heart disease, are clinically depressed. The problem is not simply that people who know they have a heart problem tend to be gloomy about it. Prospective studies have shown that, even prior to a heart attack, the presence of a major depressive disorder is a powerful predictor that the attack will take place.

When depression is combined with all other known predictive factors—ischemia, smoking, blood pressure, cholesterol levels, lifestyle—its presence accounts for a full forty percent of the predictive ability of this entire suite of factors. We must be cautious about attributing too much to the power of the mind, for the state of psychological health might have been affected by the early developing symptoms of heart disease, or some effect of the infectious disease that could have triggered the condition. But a growing body of research shows that state of the mind has a dramatic effect on physical health.

Even if, in the future, it is discovered that heart disease really is primarily caused by infectious agents, the role of clinical depression will still have to be explained. Why should depressed people be so much more susceptible to the damaging effects of these infections than people with sunnier dispositions? And what roles do genetic differences play? The interplay between visible and invisible selection, between the influences of the body and the mind, will provide fodder for scientists and sociologists for decades to come.

Adding yet another layer of complexity, the effect of these psychological factors may vary dramatically from one society to another.

Some fascinating patterns, for example, as Marmot and I discussed, have recently been observed in Japan.

In Japanese society, even though social hierarchies are extremely rigid, the differences in mortality from the highest to the lowest socioeconomic classes are reportedly very small. Indeed, they seem to be the smallest that have been measured in any industrialized society. Marmot's group is currently collaborating with Japanese sociologists to find out whether these differences in mortality are really as small as these early studies indicate.

If the early studies are borne out, the reason may be that the Japanese tend to be more accepting of class differences and take their place in the social hierarchy more willingly. The British, on the other hand, might be straining more to fight free of the chains imposed by their class structure. But if the early studies prove to be wrong, and pronounced differences in mortality—or some other indicator of stress—are discovered at the two ends of the Japanese socioeconomic scale, then the health of the Japanese, too, may turn out to be deeply affected by their position in their society's hierarchy. It also seems likely that, as younger Japanese begin to rebel against their rigid society, these stresses will increase. Socioeconomic status may not exert a differential stress, but the health effects of being out of step in Japanese society can be profound—although the gap is now closing somewhat, in the middle part of this century unmarried Japanese had life spans fifteen years shorter than married Japanese!

WHICH CIVIL SERVANTS WILL SURVIVE?

The civil servants of Whitehall, stressful though their lives may sometimes be, are not exactly dropping like flies in the corridors of power. The dry numbers summarized in these studies are difficult to reconcile with some nightmare vision of natural selection at work—Mother Nature grimly wielding her incarnadined teeth and claws while bowler-hatted civil servants scurry for cover.

Any selection that is detected by these studies would seem to be very weak. Most of the differential mortality and illness found among the Whitehall civil servants occurred in midlife, which means that it took place near or after the end of their reproductive period. But what matters most in an evolutionary sense is whether and how often the individuals occupying this social hierarchy reproduced, and what happened to their children in turn. Until recently, the Whitehall studies gave us no such information; they did not measure the effects that hierarchical position might have had earlier in life, when the civil servants were just starting out in their careers and their marriages. Did position have any effect on reproduction, or on success in parenting? And to what extent might such effects be passed from one generation to the next?

The inclusion of women in the second study should eventually allow some of these effects to be measured. A variety of gynecological problems, such as severe effects of menstruation, already appear to be strongly correlated with hierarchical position, although these results have yet to be published in detail. Again, differences are seen even between the first and second civil service categories. Do such problems, and others yet to be discovered, have an effect on reproductive fitness? The answer is still concealed in the dry lists of numbers waiting to be analyzed in Marmot's laboratory.

Even though any evolutionary connection remains elusive, the Whitehall studies have already shown that the industrialized societies of the latter part of the twentieth century are not quite as benign as they seem on the surface. Although the consequences of differences among civil service grades in Whitehall are mild, they may not be perceived as mild by people who have known nothing else. People who have relatively little else to worry about may well magnify the importance of tiny differences in status.

This tale of the civil service raises another question. It is quite clear that the Tibetans and the Quechua are being selected for the ability to survive at high altitude, and it is clear that Central Africans are being selected for resistance to malaria and other diseases. But what traits are currently being selected for or against among the civil servants of Whitehall? Are they being selected for their ability to reconcile themselves to their place in the hierarchy, so that they don't worry themselves into a state of ill health? Are genes that have a detrimental influence on mental health being selected against? Probably both, but another mental factor may be subject to selection as well.

One of the more intriguing questions on the Whitehall questionnaires was "Do you think you can reduce your risk of having a heart attack?" Seventy-five percent of the males in the highest category answered yes, while only fifty percent of those in the lowest category did. Those at the upper levels were more likely to avoid behaviors that they had learned would increase their chances of ill health, just as they were less likely to feel that their lives were buffeted by forces beyond their control. I suspect that one thing that is being selected for, even in the calm purlieus of Whitehall, is the ability to take advantage of new information and turn it to one's own benefit.

Such an amorphous ability is far less easy to define than the ability to survive at high altitude. And we have absolutely no idea of the extent to which genes might contribute to it. But throughout our history as a species, we have been subject to greater and greater onslaughts of information of all kinds. Those best able to handle this tide have benefited from it. And one day, perhaps soon, we will understand the nature of the genes that have made this possible.

HIERARCHIES IN THE JUNGLES BEYOND WHITEHALL

Although the current selection acting on the Whitehall civil servants is weak, it gives us a glimpse of the complex ways in which selection might still operate on our species.

Primate studies have already demonstrated that position in a hierarchy can have a dramatic effect on levels of stress. Studies on the baboons of Amboseli Park in Kenya by Robert Sapolsky and his co-workers show that in stable baboon hierarchies the lowest-ranked males show very high levels of endocrine indicators of stress, while the highest-ranked males have much lower levels of these indicators. But during times of upheaval, even the highest-ranking members suddenly shoot up on the stress scale. A threat to one's position in a hierarchy can be extremely stressful. It is fascinating to learn from Marmot's study of the ministry that was about to be privatized that similar effects can also be seen, however dimly, in our own species.

In addition, the baboon studies tell us something about the role of variation in a population. Sapolsky and his students found that in spite of their position, some low-level males did not show high levels of stress indicators. These males would often be able subsequently to move up in the hierarchy. They were also, it turned out, the ones most adept at stealing opportunities to copulate with females, despite the vigilance of the dominant males. These findings suggest that the ability to withstand the stress of occupying a low position on the hierarchy can increase reproductive success.

While these low-ranking males may have boosted their reproductive success by a small amount because they were alert to fleeting opportunities, recent genetic studies have shown that the high-ranking baboon males, during the fairly brief time they occupy the top ranks, reproduce most effectively. At least among baboon males, hierarchical position really does have an impact on fitness. Does this pattern extend to females as well?

Such correlations among females have been frustratingly difficult to measure, in baboons or in other animals. Although females' reproductive success is much easier to measure than that of males, the place of females in a group's hierarchy is very difficult to determine. The reason is that they usually do not show the clearly aggressive behavior of the males and tend to spend far more time alone. But despite these difficulties intriguing evidence has recently emerged that for female chimpanzees reproductive success is indeed correlated with their hierarchical position. Jane Goodall and her collaborators have patiently collected the data over a long period of time from the female members of chimpanzee groups living in Gombe National Park in Tanzania.

The researchers could find only one signal that allowed them to determine the females' social status. This is a sound called a "pant-grunt" that a submissive chimpanzee makes after a dominant one has behaved aggressively. One female would often be observed to pant-grunt at another, even though to a human observer the second female did not seem to have done anything particularly aggressive. On the assumption that an aggressive interaction had really taken place, and that this interaction, although invisible to humans, could be perceived by the lower-ranked chimpanzees, the researchers decided to use the sound as a marker.

With astonishing foresight, they had already collected just the data they needed to test this possibility. Between 1970 and 1992 they had painstakingly recorded hundreds of interactions between pairs of females. They had tallied the numbers of pant-grunts and kept track of which female was the grunter and which the gruntee. Now they were able to use this record of "invisible aggression" to unravel the female hierarchy.

Once they worked out the hierarchy, much other data fell into place. They discovered that females at the top of the hierarchy tend to have more babies, and that their babies not only have a higher chance of survival but actually mature more quickly. Thus, even though the hierarchical interactions among the females were difficult to detect, they were very important to reproductive success. And in the evolutionary long term, this is all that matters.

A final question, and a very important one, is not answered by these studies: Even if reproductive success is correlated with hierarchical position, there is no guarantee that a propensity for success will be passed down to succeeding generations. For such an evolutionary change to take place, genes that increase reproductive success must be selected for. No studies have yet shown direct evidence of such an effect.

There are, however, some hints. Frans de Waal, who has spent a lifetime observing primate behavior, found in a rhesus monkey colony in Wisconsin that mothers act in such a way as to increase the likelihood that their offspring will rise in the hierarchy. He observed a behavior called a double-hold, in which a mother enfolds her own baby and an unrelated baby together in her arms. Almost always the unrelated baby belongs to a mother with a higher rank than the double-holding mother. Similar behavior is not unknown among humans—mothers get upset when their children play with children from the wrong side of the tracks. While genes for social climbing per se are unlikely to exist, the fact that there is a hierarchy to be climbed and that there are some apes—and humans—who attempt to climb it, suggests a genetic benefit to such climbing.

It is a long way from the wilds of Gombe, or even a Wisconsin primate colony, to the calm precincts of Whitehall. Yet the social forces acting on Jane Goodall's chimpanzees and Frans de Waal's monkeys seem not so far removed from those that impinge on British civil servants. It is ironic that we know less about the effects of hierarchical position on humans than their effects on chimpanzees! Until we learn more, it would be foolish for us to rule out the possibility that natural selection is still taking place among the civil servants of Whitehall or indeed among the members of any human social hierarchy.

SIX

❖

Farewell to the Master Race

I do detest people who consider themselves superior. It
makes things so difficult for those of us who really are!

HYACINTH BUCKET, *Keeping Up Appearances*

The Whitehall studies focus on death and illness—events that we can measure fairly easily, because they happen before our eyes. But what of the events that do not happen at all? How, for example, can we measure the impact on the gene pool of a child who is never born? Just as factors that change the death rate can have an effect on evolution, so can factors that change the birth rate.

THE CHORUS OF UNBORN CHILDREN

One hallmark of primitive societies, remarked on by many anthropologists, is that almost all their members marry and have children. Many of these children die from disease, deliberate infanticide, starvation, or accidents, but in general virtually all those who survive have at least a chance to pass on their genes to the next generation. Even so, the variation in reproductive success from one human group to another can be remarkable. Among tribes in the Congo (formerly Zaire), the incidence of women without surviving children ranges from six percent among the Ngbaka to sixty-five percent among the Mbelo. It is unclear how much of this variation is due to disease, but it is likely

114

to play a substantial role: Sexually transmitted diseases, as well as malaria and trypanosomiasis, can have large effects on fertility as well as on the survival of children.

The numbers of children can vary dramatically not only from tribe to tribe but also from family to family. Extreme variation of this type can have large genetic consequences. In one study human geneticist James Neel of the University of Michigan collected data on numbers of children among families of a South American Indian group. These people, the Yanomama, live in a scattering of small villages in a heavily forested area that spans the border between Venezuela and Brazil. Their remarkable social structure was revealed to the world in a memorable book by Napoleon Chagnon, *The Fierce People*.

Neel found that even though most of the adult members of the four villages that he examined had managed to reproduce, the numbers of their children varied wildly. Further, the likelihood of survival also varied greatly from family to family. While on average only about half the children survived to reproductive age, some families were far luckier than others. Using a computer model because he did not have long-term data, he extrapolated these numbers for a further generation, to see how many grandchildren the various members of the villages would be expected to have. He was able to predict that for fifty-five percent of a typical cohort of eighty-six males in a village, none of their grandchildren would survive to adulthood. In effect, the genes of over half the cohort would disappear in the space of two generations. At least some of the genes of the other forty-five percent would survive, however, and one of the males in the cohort would probably end up with about thirty-two surviving grandchildren. His genes would almost all be passed on, and some of them would be represented many times in the population.

This real-world situation is very different from a situation governed by random chance. When I revisited Neel's data, I calculated what might have happened if the likelihood of having children were distributed equally among all males. I assumed that, just as in the real tribes, mortality among the children was fifty percent; but I specified that the deaths took place at random. Assuming that the population size remained constant from one generation to the next, I found that forty percent of the men would have no surviving grandchildren—a slightly smaller fraction than Neel predicted. But the largest number of grandchildren to be expected for any of the men would be about ten, far fewer than Neel's maximum of thirty-two.

In the real world, a lucky few of the Yanomama males will pass on their genes to subsequent generations in disproportionately large numbers. In the artificial world of my calculations, in contrast, no such extremely superfecund minority would exist.

This difference between Neel's numbers and random expectation arises because, for a variety of very nonrandom reasons, some people in a Yanomama group will be more fertile than others, and the children in some families will be more likely to survive than those in others. From Neel's data alone we cannot tell how much of these differences are due to environmental factors and how much to genetic differences among individuals. But even assuming only a small genetic contribution, the great differences in fertility and survival among the Yanomama will unquestionably have a substantial impact on their gene pool. In essence, Neel was able to show that in each generation some genes manage to pick their way through the perils that are faced by these tribespeople much more readily than others.

All these genetic events are taking place in little villages on the banks of isolated jungle rivers. The overall effect on the Yanomama of the success of some genes and the failure of others might have remained slight as long as they stayed isolated from the rest of the world. Neel has estimated that the Yanomama and their antecedents have lived in this region for a long time, perhaps a large fraction of the time since humans first arrived in South America. A time traveler able to visit a Yanomama village of a thousand years ago would probably not see many differences from the villages of today, and certainly the appearance of the people would be little different.

Now, as the twentieth century is beginning to impinge on their ancient way of life, that world is vanishing. Many Yanomama have died from introduced diseases, and others have moved away to the cities and shantytowns that are springing up everywhere in the Amazon basin. As a recognizable people, they will be gone in an evolutionary moment, even though some of their genes will undoubtedly persist. And as the genes of the Yanomama are catapulted into a new environmental context, they will survive or disappear according to very different rules.

Theirs is only a tiny part of the enormous genetic upheaval that is taking place throughout the world. The effects of all these selective processes, especially the effects of children who were never born, can have dramatic and sometimes unexpected consequences on this larger canvas. The gene pools of entire continents of people are being altered.

THE CHANGING FACE OF EUROPE

Two kinds of genetic change are taking place globally, one obvious and one far less so. The obvious change is the rapid mixing of gene pools that have remained separate for millennia. The results are seen most dramatically in people who live in major cities throughout the world. The London that my parents knew, for example, has undergone enormous alteration in the space of two generations, and the rate of change is accelerating.

The second kind of change is one that receives its impetus from this mixing of different gene pools. As we saw in Chapter 1, mixing can alter allele frequencies and produce new gene combinations. But real evolutionary change will only happen when these new gene combinations are sorted out by natural selection, so that the frequencies of the alleles in these new mixed populations begin to shift.

Europe provides some particularly dramatic examples of both kinds of change, and of how traumatic the changes can be. In recent history, various groups who imagined themselves to be indigenous and permanent have had great difficulty coming to terms with the fact that, as genetically separated populations, they are doomed to extinction.

Two generations ago the gene pools of Europe were the jealous preserves of xenophobic demagogues. Attempts to "purify" them led to some of the most ghastly events in the history of our species. At the present time this "purification" process, with all its attendant hatred and misery, continues in various sad fragments of the former Yugoslavia. But despite the efforts of Hitler and his successor Karadzic, and in one of the great ironies of history, the current populations of Europe will soon be altered beyond recognition.

New mixed populations can swamp and overwhelm old ones for many reasons. New technologies, such as agricultural innovation or rapid transportation, may drive the process. In a collision of cultures the vibrant new will tend to outstrip the fossilized and hidebound old. Psychological factors can also play a central role. When sociologists have examined the effects of mental attitude on demographic trends, they have discovered them to be profound. And demographic trends are, after all, the very stuff of the evolutionary process.

One important finding that has emerged from such studies is that social attitudes can have a substantial effect on birth rates. In an astonishing reversal of historical trends, the native populations of many European countries are currently declining in numbers. In European Russia, from 1989 to 1993, the birth rate dropped by thirty-five percent, and in eastern Germany during the same period, the rate plunged an astounding sixty percent. On average, if a population is to remain at steady state, each woman must have about 2.2 children. In eastern Germany, assuming current trends are not reversed, the average number of children born to each woman during her lifetime will be less than one.

At the same time death rates have soared in Eastern Europe, putting this region at odds with every other part of the world. Even in Africa, in spite of warfare and economic and epidemiological disasters, death rates have by and large fallen.

The role of psychology in this trend is illustrated vividly by the gap between the life expectancies of Russian men and women. Female babies born in Russia at the present time are expected to live 73.2 years, compared with an expectation for white female babies in the United States of 79.6 years. But male babies born in Russia will only live 58 years, while white male babies in the United States live an average of 73.4 years. The gap between Russian men and women is more than twice as large as the gap between American men and women, and indeed Russian men live about as long as men in Liberia.

Why the huge difference between Russians and Americans? Many causes have been suggested, from the declining quality of health care in the former Soviet Union to the effects of widespread and extreme environmental pollution. The disproportionate effect on Russian males can in part be traced to high levels of binge drinking and smoking. But there is another important cause, a psychological one. A combination of stress, discouragement, and helplessness—of the sort that seems to affect the lower echelons of Whitehall civil servants but greatly magnified—becomes clearly visible when we examine these Russian numbers more carefully.

Although I compared the Russian life-expectancy statistics with those for white Americans, it might be more informative to compare them with the statistics for black Americans. Life expectancy for black females born in the United States in 1995 is 74.0 years, but for black males it is only 65.4 years. These numbers, and the huge spread between them, are remarkably similar to the equivalent numbers for whites living in Russia.

The psychological situations faced by these two superficially very different populations may have similarities as well. In the United States the self-image of black males is under severe assault, chiefly because of the pervasiveness of racial discrimination. The self-image of Russian males is under equally severe assault, as they find that their society is no longer either stable or a superpower but has been demoted to a marginal third-world entity. Females, for a variety of reasons, seem to be less severely affected by such assaults on self-image.

These birth- and death-rate trends have already had an enormous impact on the Russian population, which in 1995 alone actually decreased by 900,000. If the trends continue, then by the year 2030 the population of Russia will, in spite of increasing immigration from other republics, have fallen from 148 million to 123 million. And the ethnic mix will have changed dramatically.

Eastern Europe and Russia present us with a gigantic experiment in which human misery and a decline in fertility are obviously correlated, although just how much the first causes the second remains uncertain. The effects of psychological attitudes on reproduction are notoriously difficult to measure. It has been repeatedly suggested, by analogy with John Calhoun's famous experiments on rats in the early 1960s, that stress due to crowding should decrease birth rates in people. Psychologist Jonathan Freedman, in summarizing many studies that measured the effects of crowding on the attitudes of individuals, found that these effects can range from positive to very negative depending on the design of the experiment. But one generalization does emerge: If people have the perception that their surroundings are both crowded and unpleasant, they acquire strikingly negative attitudes toward life in general.

In another few decades a sunnier economic outlook in eastern Germany and Russia, and the disappearance of the older generation that had been used to life under Communism, will likely change things again. Birth rates will rise, although they are unlikely to reach the levels prevalent before the collapse of Communism.

The demographic upheavals in Eastern Europe are an extreme manifestation of a swift and continuing demographic transition throughout the continent. Italy, in spite of the pronatalist stand of the Catholic Church, now has the lowest birth rate in Western Europe. At 1.3 children per woman, it is far below the 2.2 required for replacement. Prosperous and politically stable western Germany has a rate that is almost as low. The united Germany is now predicted, if current trends remain unchanged, to experience a population decline from 80 million today to 55 million by 2050. This prediction has surprised demographers, who have assumed that as prosperity and education levels increase, populations tend to come into equilibrium rather than plunge.

Even the general public seems to have expected an equilibrium, not a decline. In 1947, 73 percent of French people surveyed thought that the population of France should increase, while 22 percent thought it should stay the same and only one percent thought it should decrease. By 1974 the upbeat pronatalists had declined to only 23 percent, those who thought the population should stay the same had risen to 63 percent, and those who thought their numbers should decrease had risen to 10 percent. The decline in birth rates has actually gone far beyond the expectations of the people experiencing the decline.

Demographic transitions, in which declining birth rates follow upon declining death rates, are now taking place in many parts of the world. Eastern Europe's transitions are particularly striking, since death rates are rising as birth rates are falling. Before these calamitous changes, the most rapid demographic transitions for which we have records took place in Korea and Taiwan. In Korea, from 1960 to 1974, the crude birth rate was cut almost in half, while similar declines were seen in Taiwan.

The reasons for these declines are numerous, but the transitions undergone by these countries are very different from the widespread economic upheaval and discouragement that have had such an impact on Eastern Europe. Korea and Taiwan appear to have reached a kind of societal consensus, for all classes and educational levels have been affected by the decline.

This pattern is very different from that seen in less-developed countries that have less economic equality. There any decline in birth rate tends to be concentrated in the well-off or educated classes. In sum, the reasons for birth declines are as numerous as social structures and local histories.

Worldwide, the demographic change over the last few decades has been dramatic. Population growth has a kind of inertia—the huge numbers of people currently moving into their reproductive years ensure that world population will continue to grow overall well into the next century. The peak numbers that will be reached, however, are continually being revised downward. In 1960 the median United Nations estimate for global population by 2050 was twelve billion. Now the median 1997 prediction is that it will reach a mere 9.1 billion by the middle of the next century, and this estimate seems certain to drop further. This number is still huge and frightening, but it is three billion less than was predicted less than forty years ago.

Most tellingly, by the year 2015, at least two-thirds of the world's population will live in countries in which the number of children per woman has dropped below the replacement level of 2.2. Malthusian terrors may still await us, but as a species we seem to be responding with commendable rapidity to the new pressures of an overcrowded world.

These precipitous drops in birth rate have caught everyone by surprise—or almost everyone. In 1952 Charles Galton Darwin, a grandson of Charles Darwin, wrote a book entitled *The Next Million Years*, in which he predicted that in the future people will move from the developing world into the first world, as a result of the population vacuum caused by the demographic transition there. His prediction now appears to be coming true in Europe. In spite of intense xenophobic reactions in many countries, migration from the less developed world into Europe is increasing rapidly. Almost ten percent of the population of Germany is now foreign-born.

Such changes are happening far more quickly in some cities and countries than in others. London and Paris will soon be as polyglot as Los Angeles, but Madrid and Stockholm are likely to lag much further behind. Even the melting pot is defined differently in different places. I recently met an Irish archaeologist and over dinner regaled him with tales of this European transformation. "I know just what you mean!" he exclaimed. "One of my coworkers back in Dublin has just married"—and here he paused for effect—"a Frenchwoman!"

AN EVOLUTIONARY EYE-BLINK

My collaborator Pascal Gagneux, a student of chimpanzee behavior, tells me that when he was a child in Switzerland, all his relatives and indeed everyone he knew was white. Now, each time he revisits his native country, he finds that the ethnic mix has changed dramatically. So exotic is this mixture becoming that children of Tibetan refugees are now joining the Swiss army in substantial numbers. As in most of the rest of Europe, the population of Switzerland is altering at unprecedented speed.

A few centuries from now—an eye-blink in evolutionary terms—the peoples who occupy the European peninsula will have undergone a huge change. We can bid farewell, at least in their present forms, to the "master races"—and the new, dynamic mixtures of peoples emerging throughout the planet will likely not mourn their passing.

Indeed, "master races" seem to be on the decline everywhere around the world. In South Africa in 1951, people officially classified as white by the apartheid regime made up a fifth of the total population outside the "black homelands." By the year 2020 they will make up one-eleventh of the total. This demographic shift helped bring about the demise of that unlamented regime.

Other groups that have acquired economic power and high levels of education, and thus constitute what we can call "master races," are declining almost as rapidly. Sometimes, by an accident of history or geography, these groups are diminishing in numbers but immigrants are not yet altering their gene pool. The Japanese are a notable example.

Japanese birth rates began to decline in the 1920s, then shot up just before and during World War II, because a government desperate for soldiers instituted an unrelenting pronatalist campaign, including severe penalties for abortions. In 1947 the Japanese lifetime birthrate was 4.5 children per woman. But the decline soon resumed, aided by the legalization of abortion in 1949, and by 1957 it had plunged to 2.0. Starting in 1974, a further decline began, and Japanese birthrates are now well below replacement.

Dramatic demographic shifts have accentuated the effects of the decline. The Japanese countryside has emptied out as people have moved in huge numbers to the cities. There, overcrowding makes it virtually impossible to have large numbers of children. More than twenty desperate towns and villages, faced with the prospect of becoming ghosts of their former selves, have posted rewards for large families. The richest prizes are being offered by the government of Hachijojima, an island southeast of Tokyo: a graduated scale ranging from half a million yen for a third child up to a dazzling 2 million (about $20,000) for a fifth. During the first year after the reward system was begun, five couples made claims—not enough, however, to have much of a demographic impact.

The Japanese have one of the world's longest life expectancies and have reduced infant mortality to the lowest rate found anywhere in the world. Married women still have an average of about two children, but they are getting married in smaller and smaller numbers. In 1975, 21 percent of women between the ages of 25 and 29 had never been married. By 1990 that proportion had doubled, to 40 percent. Women complain bitterly that men are too exhausted by their endless workdays to manage the additional strenuous activity of baby-making. And once the babies have arrived, men do nothing to help their wives, even though most of the wives are also holding down demanding jobs.

If these trends continue, a population vacuum will emerge in Japan. Despite the isolation of Japanese society, it will be necessary for industries to begin importing guest workers. Then, as has already happened in Europe, the transformation of the gene pool in Japan will begin in earnest.

In my own state of California, people who call themselves white on census forms will become a minority by the year 2000, and the Census Bureau estimates conservatively that whites will be in a minority in the entire United States by 2050. My own hunch is that these demographic trends will accelerate, as they are doing in Europe, and whites will be demoted to minority status much more quickly than that.

A good deal of the demographic change will be driven by intermarriage. My own grandchildren, when (as I devoutly hope) they eventually make their appearance, will be one quarter northern European, one quarter Chinese, and the rest a mix of southern European and Peruvian Indian. In the United States the number of children classified as being of more than one race has risen from fewer than half a million to more than two million from 1970 to 1993. The rate of racial mixing continues to accelerate.

Extrapolating demographic trends over the long term is dangerous in the extreme. Declining population sizes have as many different causes as there are populations undergoing decline. Moreover, these trends will often be locally reversed, at least temporarily. But two important points can be made that have implications for the future evolution of our species.

First, the average member of our species a century or two from now will be heterozygous at more different genes than the average person is today. Isolated islands of relative genetic homogeneity that have built up over tens or hundreds of thousands of years will have disappeared.

Second, even though recognizable members of the various "master races" will be far less common than they are today, their genes will not disappear as well. Rather, their genetic legacy will still be there, combining and recombining in ways that the fierce xenophobes of the past could never have imagined.

Is all this change evolutionary in nature? You bet it is—rapid and fundamental changes in the gene pool do not require terrible plagues or the slaughter of war.

Our current population size is unsustainable—a millennium from now, the world population will doubtless be far smaller than it is today. Much—and I hope most—of this shrinkage will be the result of individual decisions to have fewer children or none at all. Although we cannot predict what the effect on the gene pool of the survivors will be, it is likely to be profound.

In this first part of this book, we have looked at the wide range of evolutionary pressures acting on us, some obvious and some more subtle. Physical factors in the environment, such as disease, are playing a smaller role, though they are not declining as much as optimistic scientists had suggested a few years ago. Social and psychological pressures are increasing, and their nature is changing rapidly.

Some authorities contend that the benefits conferred by our complex culture have somehow brought our pell-mell evolution to a halt. If they are right, it would be an event unparalleled in our long history. If the evolutionary brakes have been applied so strongly, the smell of burning rubber should be everywhere!

PART II

❖

Our Stormy Evolutionary History

SEVEN

❖

The Road We Did Not Take

As recently as 35,000 years ago western Europe was
still occupied by Neandertals, primitive beings for
whom art and progress scarcely existed.
JARED DIAMOND, "The Great Leap Forward" (1989)

The rolling brown and olive-green hills of the northern Spanish province of Castile and Léon have seen a great deal of history. In medieval
times millions of pilgrims tramped the length of the province on their
way to the great cathedral town of Santiago de Compostela. The pilgrims believed that, in the ninth century, Saint James himself had arisen
from his grave and beaten back the Moors in a great battle. During
succeeding centuries the pilgrimage route to the site was marked with
thriving towns and astonishing cathedrals, standing as testaments to
the overwhelming force of faith.

But the history of the region dates back much, much further. These
same hills have witnessed the triumphs and tragedies of an ancient group
of western European people who had evolved their own unique characteristics. Their story was played out over a very long span of time,
starting with their arrival in the region well over a million years ago and
ending with their final disappearance less than thirty thousand years ago.

The remains of the last people of this lineage were first discov-
ered a century and a half ago. We call them the Neandertals, after the
Neander valley in Germany, where traces of them were first found.
The Neandertals' forebears were ancient tribes whose real names have

127

long been lost to history but whom we now call pre-Neandertals. Some of the most vivid evidence for their history has been found among the Spanish hills.

New fossil finds are casting fresh light on this remarkable story, which I recount here for three reasons.

First, these finds show that the roots of our own species go a long way back in time, much further back than most people think. Human beings did not suddenly appear on the planet a few tens of thousands of years ago, in the form of modern people, different from anybody who had gone before. We did not suddenly acquire the ability to decorate the cave walls with pictures and patterns and make elegant artifacts of wood and bone. Our emergence, like the parallel emergence of the Neandertals, took far longer.

Second, the same strong forces that have driven our evolution also drove the evolution of the Neandertals. We are not unique. Both we and they were caught up in a period of ever-accelerating evolution.

Third, and recent news stories to the contrary, the genetic differences between us and the Neandertals are not profound. Genetic differences among present-day human groups are actually very slight, primarily involving different frequencies of alleles of various genes that all human groups hold in common. We saw in Chapter 1 that if populations differ not in their genes but in the frequencies of alleles of their genes, the effects can be substantial. Selective pressures or chance events can easily change such populations by shifting the frequencies of the alleles. The Neandertals, while somewhat different from us, were not so far away as to have a different collection of genes; rather, they were primarily distinguishable from us by different allele frequencies.

Many anthropologists ignore this evidence and continue to consider Neandertals to be not quite human. But I think it is scientifically far more valid to hold up their story as a mirror of our own, and to learn as much as we can from it.

The emerging story of the pre-Neandertals of northern Spain was uncovered through the patience and determination of the Spanish archaeologists Emiliano Aguirre, Eudald Carbonell, Juan Luis Arsuaga, and their colleagues, who persisted for years in excavating sites that everyone else told them were hopelessly unpromising.

The sites lie quite near the ancient cathedral town of Burgos, with its charming gardenlike moat and immense carved city gate. To the east of Burgos the undistinguished village of Atapuerca has given its name to a low range of limestone hills. The limestone was originally laid down in a warm tropical sea during the last part of the age of dinosaurs. Much later, water seeping from above carved out a vast honeycomb of caves.

Very little of this region has been excavated by archaeologists, so it is particularly striking that two out of the three sites that have been examined intensively have yielded important human fossils. For such fossils to be so plentiful, the whole region must once have supported large populations.

Superficially, the world inhabited by these people would not have been much different from what we see today. Grazing and farming have now transformed the nearby lowlands, but the hills are still covered with scrub oak and wild rosebushes. Grasshoppers make a deafening noise. Pollen grains found in the digs suggest that the vegetation has not changed greatly over the last million years—the peoples who lived here during all that time probably looked out over a very similar landscape, sometimes a little more heavily forested, sometimes a little less. The big difference was in the animals. In earlier times many different animals flourished here, including elephants, boars, cave bears, and rhinoceros; deer flourished in the region, making it ideal for hunting. The last of the remaining animals were hunted to near-extinction by the nobles of Castile during medieval times.

A RAILWAY TO NOWHERE

During the latter part of the last century a deep and narrow railway cutting known as the Gran Dolina was sliced through the nearby hills. The railway was used only briefly before it was abandoned. Today it is one of the most dramatic archaeological sites in the area.

At several points in the cutting, the workers had exposed old caves, actually pockets in the limestone that, over the millennia, had been completely filled by detritus washed in from above. The detritus is made up of compressed orange earth, some of which spilled out into the cutting.

The archaeologists excavated down through these layers, looking for information about the hundreds of millennia during which the caverns had slowly filled up. They excavated two of these caverns, each filled with material some eighteen meters thick. The first, discouragingly, yielded nothing. The second also seemed empty, except for a few possible stone tools. But, in the early 1990s, the excavators took a chance. Leaping ahead of their normally methodical investigation, they tunneled a one-meter-square hole down toward the bottom of the deposit. In 1994, their gamble paid off, as they discovered stone tools and fragmentary human remains at the nine-meter level.

Dating the find might have been a problem, but luckily it lay beneath a layer of deposit that marks a sudden reversal of the earth's magnetic field. The switch is one of many that have taken place periodically throughout our planet's history. Traces of this particular reversal, known as Matuyama-Brunhes, have been found in many other places, allowing the event to be dated with precision at 780,000 years before the present. The Gran Dolina bones and artifacts are therefore older than that, perhaps substantially older. Indeed, they are the oldest fossil human bones, by at least three hundred thousand years, that have ever been found in Europe.

At the Natural History Museum in Madrid, José Bermudez de Castro showed me the finds. The stone tools are very primitive, of a type known as Oldowan. The bones themselves are few, including some isolated teeth, a fragment of an upper jaw, and some bits of lower mandible that include molars. They are, however, remarkably well preserved—it is still possible to see the gum line on some of the teeth.

The most important find is a piece of facial bone, including the lower rim of an eye socket and enough of the upper jaw to determine that it belonged to a child. The angle of the facial bone below the eye socket is surprisingly vertical, and as a result the child who owned this face has been put by its discoverers into a separate human species, *Homo antecessor*. This new species has been proposed to be an ancestor of both Neandertals and modern humans. But the child's modern-looking face could also simply be the result of the fact that the skull is immature—even the skull of an immature chimpanzee has much in common with that of a human child. My own guess is that this child, living the better part of a million years ago in these remote hills, was already following the separate evolutionary path that led not to modern humans but to Neandertals.

The fact that so few artifacts and remains have been found after so much excavation suggests that the people of the area probably did not spend much time in the caves but rather lived and hunted out in the open. Little can be inferred about how the child's remains, along with a few fragments of an adult cranium, came to be there. Cut marks on the bones suggest that they might have been butchered and eaten, and their remains thrown into the cave afterward.

The discoveries in the Gran Dolina have already pushed the first appearance of hominids in Europe much further back in time than might have been imagined just a few short years ago. In addition to this discovery at the nine-meter level, at least two deeper layers—which have yet to be investigated thoroughly—are known to contain stone tools tentatively dated to well over a million years ago. It is likely that further excavations will yield even older traces of human activity.

Hominids first ventured out of Africa into Asia, and perhaps into Eastern Europe, at least 2.5 million years ago. It seems unlikely that it took them a million years to travel as far west as the Sierra de Atapuerca. Either at this site or elsewhere in this honeycomb of caves, bones will certainly be discovered that will help fill in that million-year gap

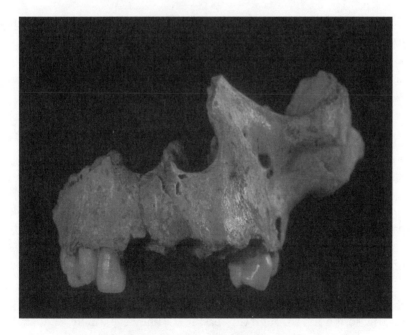

Figure 7–1. The face of the Gran Dolina child. In spite of its great age, it is surprisingly narrow and vertical.

between the first ancient peoples to leave Africa and the people of the Gran Dolina.

These excavations provide only part of the complex story of early human history in Europe. There is much evidence that hominids occupied other parts of the continent, beginning at least half a million years ago.

These other finds, like those at Gran Dolina, are fragmentary, consisting of isolated jaws, crania, and leg bones. They have turned up at sites scattered over a huge area that extends from the Pyrenees to southern England and northern Germany and east into Central Europe and Greece. The finds are puzzlingly variable, showing some features that are Neandertal-like, others that are more primitive, and still others that defy categorization altogether. Yet their variability may be no more significant than the variability found among human populations today. Study of these early remains, like that of the Gran Dolina people, has been hampered by their very fragmentary nature. For example, one species, *Homo heidelbergensis*, which probably lived half a million years ago, has been named entirely on the basis of a single lower jaw.

The fragmentary nature of these finds has been the curse of paleontologists, who have been forced to infer complex evolutionary stories from a few bits of bone. For this reason a second great discovery that has emerged from the caves of the Sierra de Atapuerca is dazzling. It allows archaeologists to examine ancient Europeans in enough detail to get a good idea of what they must have been like as a group, not as a few broken bits of one or two individuals.

THE CAVERN OF BONES

A few hundred meters from the Gran Dolina railway cut, a winding path leads to a large cave entrance under an overhanging vertical cliff of limestone. Until it was recently closed off by a metal gate, generations of explorers had found bones of cave bears in abundance in the thick wet mud deposits on the cave's floor.

The cave consists of a series of large chambers, separated by narrow scrambles through tiny passages. They eventually lead, after some five hundred meters, to a heart-stopping drop-off. At the bottom of this hole, thirteen meters deep, is a cramped, claustrophobic, sloping chamber with a floor only a few meters square. Over the years many cave explorers climbed down into this dark hole, trampling the floor and leaving garbage behind. It became known as the Sima de los Huesos, the cavern of bones, from the bear bones that were found there.

One day in 1976 Trinidad Torres, a professor of mining, and a spelunker named Carlos Puch had descended together into this uttermost cave in search of cave bear bones. Suddenly Puch made a remarkable discovery—in the midst of the cave bear bones was a human mandible! As soon as the two emerged, they took their find to Emiliano Aguirre, an expert on fossils, who realized its importance.

Two years were to pass before Aguirre managed to get a team together to explore the cave further. They soon found that excavation was a daunting task—part of the cave's cramped floor was covered with huge limestone blocks that had to be removed. This task, and the excavation of the overburden of trampled and damaged sediments, took several years. The excavators had to fill backpacks with rock, then climb and crawl through the narrow passages back to the surface. A couple of these grueling trips a day were the most that could be managed. So deep was the cave that if more than two or three people worked there simultaneously, they soon began to run out of oxygen and were forced to climb out.

Finally, after they removed three tons of material, undisturbed sediments were revealed. Excitement was growing, for sieving of the overburden had already revealed many fragments of human bones. The degree of preservation was extraordinary—even some tiny bones of the middle ear had been found. This little cave clearly contained many disarticulated, badly fragmented, but complete skeletons.

Some of the undisturbed sediments were found under a layer of stalagmitic limestone, which could be dated at about 300,000 years old. The bones are therefore at least that old, which makes them approximately half as old as the fragments that would later be found in the excavations of the nearby Gran Dolina.

It is now estimated that the complete skeletons in the Sima de los Huesos number more than a hundred and that it will take decades to finish the excavation. The way they are piled together in a tangled heap indicates that these people perished over a very short span of time. How did they come to be there, so far below ground, in such a tiny place?

We simply do not know. Perhaps they died elsewhere and their bodies were thrown down that thirteen-meter shaft. Or perhaps they wandered in through a lower entrance that has since disappeared—such an entrance may have existed at the time. If they did, a terrible fate awaited them. They might have been overcome by lack of oxygen, as nearly happened to some of the later excavators. They might have been trapped behind a rockfall. Or they might have been herded or thrown in there by some conquering tribe and left to die, a grim early case of genocide lost to history. Clues to what happened to them may eventually be found among the undisturbed layers.

Once those undisturbed layers were reached, the pace of excavation slowed, for the position of every bone had to be measured carefully. To make things more difficult, the bones had been wet for hundreds of thousands of years and were now so soft and fragile that a touch could damage them. Whole chunks of the floor of the cave had to be dug out and carefully dried in order to allow the bones to harden. A shaft was drilled down from the surface, through which material could be hauled up. As a bonus, precious oxygen could now percolate down the shaft, so the excavators far below no longer had to put their brain cells at risk from repeated oxygen deprivation.

Finally the years of work paid off. On July 7, 1992, a complete cranium was found, and others were soon discovered nearby. Freeing the blocks containing the crania and bringing them to the surface took two weeks of exquisite effort. The skulls could not be matched to any of the lower jaws that had been brought up earlier, although a year later a jaw was found nearby that matched the largest and best-preserved skull.

Over the years I have seen many different fossil finds from sites in Africa, Europe, and Asia, but nothing prepared me for the impact of the bones of the Sima de los Huesos. Their sheer abundance is staggering. With a flourish José Bermudez de Castro opened a drawer in a cabinet at the museum to reveal a dozen superbly preserved lower jaws. At the nearby Universidad Complutense, Juan Luis Arsuaga showed me the skulls, which were also remarkably well preserved. One showed signs of a massive upper jaw infection, which might have been the cause of death.

Both of the best-preserved skulls are adult, but their crania are very different. One had a large brain, comparable in size to those of modern humans, though not quite as large as those of the much later Neandertals. The other had a much smaller brain, so small that it actually fell within the range of our even earlier ancestor *Homo erectus*. If so huge a variation in brain size had been found in different excavations, the fossils could easily have been attributed to different species. We know, however, that they belonged to the same tribe. It is likely that, as the excavation proceeds, even more variation will be discovered. Even though this variation may seem extreme, a similar range of variation sometimes appears within human groups today.

Such variation cannot be investigated among any of the other early European fossil finds, since they usually consist of a few isolated bones. Indeed, so abundant are the fossils of the Sima de los Huesos that Arsuaga and his coworkers have been able to show that the differences in body size between males and females are about the same as in present-day human populations. These people were like us in many ways.

But not in all ways. Like the skulls of the much later Neandertals and unlike our own, the skulls in the Cavern of Bones have prominent browridges and rather forward-thrusting faces. These features were, however, not as pronounced as they were later to become in the Neandertals, especially those that lived in the west of Europe, nor were the bones of their skeletons as heavy as those of the Neandertals. (See Figure 7–2.)

The differences in size do not extend to the jaws and teeth, which are remarkably uniform. The mandibles are massive and very like those of Neandertals—and very like that of Heidelberg man. Neandertals, like modern Eskimos, used their front teeth as pliers, gripping objects with such force that the front teeth wore down more quickly than the molars. The people of the Sima de los Huesos were doing the same, for without exception the front teeth of their jaws are badly worn down while the crowns of the molars are relatively intact.

The skulls of the Sima de los Huesos differ from those of the western Neandertals in one striking respect, however. Neandertal skulls, as Jean-Jacques Hublin had earlier shown me in a basement room of the Musée de l'Homme in Paris, have a rounded, swollen appearance, almost as if the back of the head has been blown up like a balloon. Such an extreme "occipital boss" is seen in no other hominid fossils from any other era. It is certainly not present in the skulls of the Sima de los Huesos. Rather, when observed from the rear, those

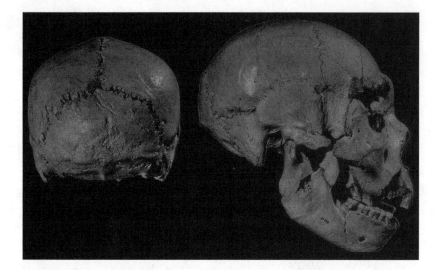

Figure 7–2. The best preserved Sima de los Huesos skull, from the rear and from the side. Note the pentagonal shape of the skull from the rear.

skulls have a pentagonal appearance, shaped by the clearly visible angles of the parietal and temporal bones on their sides. When viewed from that vantage, apart from their low cranial vault they could be mistaken for modern human skulls. (See Figures 7–2 and 7–3.)

Nevertheless, the bones that Arsuaga and Bermudez de Castro showed me convinced me that the people of the Sima de los Huesos were firmly in the lineage that eventually led to the western Neandertals. The only important differences between them were the enlarged and swollen braincase at the rear of the Neandertals' skulls, and the fact that the Neandertals tended to have more robust skeletons. The pattern of tooth wear, the shape of the jaws, and the size and shape of the projecting browridges at Sima de los Huesos were all clearly like those of the much later Neandertals.

A SEPARATE LINEAGE

Thanks to these Spanish discoveries, our picture of the Neandertal lineage is growing more complete. The Neandertals' history has been dramatically lengthened and has been revealed to be more complex than scientists had previously imagined.

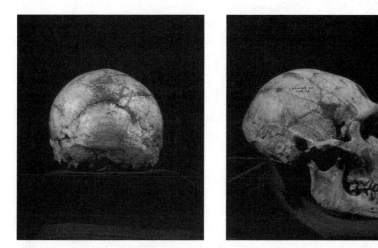

Figure 7–3. On the left, a rear view, and on the right, a side view of a typical late Neandertal skull, from La Ferrassie. Note the swollen appearance of the skull from the rear, very different from that of the skulls found in the Sima de los Huesos.

The child of the Gran Dolina probably lived relatively near the base of the branch that led to the western Neandertals. It must have been the descendant of some group of people who, more than a million years ago, wandered as far west as it is possible to go in Southern Europe. There they became isolated from the evolutionary events happening elsewhere, and started on their own separate evolutionary path.

Three or four hundred thousand years passed between the time of that child and the time of the people of the Sima de los Huesos. These later people, like other peoples living in Western Europe at the same time, had large brains. They probably had an advanced technology, although no tools were discovered with the bones: the evidence comes from a remarkable archaeological find much farther to the north.

This stunning discovery was made in early 1997, by Hartmut Thieme of the Institute for Historic Site Preservation in Hanover, Germany. For years he and his coworkers had been monitoring excavations at an open-cast brown coal mine near the city. The gigantic digging machines used at the mine often turn up important sites and artifacts, and Thieme and his colleagues watch for such events. When something is found, they must react quickly in order to stop the machines and examine the site in detail. As soon as they are finished, the machines start up again.

The site that was uncovered in 1997 was particularly remarkable. It yielded three long, heavy, and painstakingly carved wooden spears, along with a throwing-stick and a variety of stone tools. These wooden artifacts would normally have long since rotted away, but they had been preserved in peat. Numerous bone fragments were found nearby, including the remains of ten horses with cut-marks on their bones, showing that the site had been used for the butchery of animals. Other signs of human activity, including a primitive hearth, were also uncovered.

These finds would have been unremarkable if they had only been a few thousand years old, but to everybody's surprise they were dated very clearly to about four hundred thousand years ago. This astonishing date makes the spears the oldest wooden artifacts ever found. The people who made them were approximate contemporaries of the people of Sima de los Huesos.

This discovery triples the period of time for which we have evidence of systematic hunting among our ancestors or near-ancestors. The spears are not simply pointed sticks but are very sophisticated, indicating a long spear-making tradition that went back even further in time. They were undoubtedly meant to be thrown and were quite capable of killing big game. Their points had been carved from the base of the sapling's trunk, where the wood is hardest. And just as with a modern javelin, they are carefully balanced so that their centers of gravity are about a third of the way back from the point. They provide unequivocal evidence that at least some of the people of western Europe were technologically well advanced at the time the tribe at Sima de los Huesos met its fate.

We are constrained to infer the level of any really ancient culture from the few artifacts that happen to have been preserved. Stone tools survive very well, while wooden implements do not. It now appears that their primitive stone tools may have fooled us into selling the ancestors of the Neandertals short. These people were much smarter, for much longer, than we had imagined. We will probably not find advanced wood-based technology in the Atapuerca cave deposits, but the Sima de los Huesos people may very well have used such weapons to hunt in the surrounding hills and plains.

These lucky finds from 300,000 to 400,000 years ago give us remarkable clues about these peoples. But a frustrating gap appears in the fossil record, from the time of the Sima de los Huesos people down to about 100,000 years ago. Only then do we find the first evidence for true Neandertals.

The most extreme types of Neandertal remains have been found in caves ranging from southern Spain to northern Germany. Less extreme types have turned up throughout Eastern Europe, and as far east as Israel, the Crimea, and Iran. Erik Trinkaus, an expert on the later Neandertals, suspects that these less extreme types were already mixing genetically with other types of humans who were more like ourselves. The western Neandertals, however, because they lived in such remote and isolated regions, remained separate.

ANCIENT DNA

The great age of the western Neandertal lineage has recently been demonstrated directly, through a daring exploit by Matthias Krings, a graduate student at the University of Munich. Krings disobeyed the first rule followed by students who want to get a Ph.D: He embarked on a project that had very little chance of succeeding.

Was it, Krings wondered, possible to get intact DNA out of a Neandertal bone and sequence it? Svante Pääbo, his adviser, was the world's leading expert on extracting old DNA from museum specimens and fossils, but he was very doubtful that it would work.

The DNA that Krings set out to find is carried in structures inside our cells known as mitochondria. These structures, tiny but absolutely essential, are the factories that produce the huge amounts of energy-rich compounds that we need for our growth and development.

Mitochondria are not an integral part of our cells but are actually the remote descendants of creatures that were originally quite alien to us. Their presence in such intimate association with our own bodies is the result of a remarkable evolutionary event that took place among some of our single-celled ancestors about two billion years ago.

At that time a species of parasitic bacterium attacked and entered our ancestors' cells—probably producing a nasty disease in the process. Over time, however, just as can happen with present-day diseases, host and pathogen gradually reached an accommodation. The result was beneficial to both sides: the bacteria found a safe place to live, passing from one generation to the next inside their hosts' cells, while at the same time they brought a great gift.

Unlike the cells they were attacking, the bacteria were able to utilize their food fully, extracting most of the available energy from their diet. They did so by using oxygen to burn the food in a highly controlled way. When they had evolved to the point of invading the cells of their hosts without killing them, the hosts also began to benefit from all this new energy, taking evolutionary pathways that had previously been closed to them. They could, for example, grow in size and complexity, and eventually some of them could become clever multicellular organisms like us.

We think of ourselves as oxygen breathers, but in fact we are not. Without our mitochondria, we would die instantly. That is why cyanide is fatal—it destroys this essential function of our mitochondria, while leaving the rest of our bodies quite unharmed. (Bacteria that do not breathe oxygen are unaffected by cyanide; some of them actually use it as an energy source.)

By now, the mitochondria have long since degenerated into nothing more than tiny energy factories, and most of their genes have been lost or transferred to the chromosomes of their hosts. But for reasons that are not understood, a few genes have been left behind, carried on a tiny bit of DNA.

Mitochondria are not transmitted by sperm, which means that they can be passed down only from mother to daughter through the egg. Nor do they seem to recombine with each other, with the result that each mitochondrial chromosome sends intact copies of itself to the next generation. But these chromosomes, like any other pieces of DNA, do acquire mutations. Over time they gradually diverge from each other, like sentences copied by scribes who make the occasional mistake. Krings realized that if he could isolate Neandertal DNA and compare it with that of present-day humans, it would tell him a great deal about the differences between the Neandertal lineage and ours.

As Krings knew, mitochondrial DNA has one other overwhelming advantage. Most of our genes are present in two copies in each cell, one from our mothers and one from our fathers. But mitochondrial genes are present in hundreds or even thousands of copies, because there are many mitochondria in each cell and each mitochondrion may have many chromosomes. Because the DNA in fossil bones breaks down over time, it is much easier to find a given piece of ancient mitochondrial DNA that is still intact than it is to find a given piece of DNA from the nuclear chromosomes.

Krings began by examining the very first Neandertal ever discovered, a partial skeleton found by quarry workers in the Neander valley in 1856. The bones had been buried in freezing cold mud for thirty thousand years. Then they had been transferred to a museum display, where they had sat quietly—except of course for the innumerable times during which they had been brought out and handled by scientists and the simply curious. Tests showed that their protein had suffered very little damage, so perhaps some of the DNA had survived intact as well.

If any of the original DNA were still left in the remains of the bone cells, Krings knew, it would probably have been broken up into fairly short pieces after all this time. But the many people who had handled the bones must have left bits of their DNA behind, and this DNA would be in much better shape. Somehow this contaminating DNA had to be distinguished from the ancient DNA of the Neandertal.

To make things harder, most of the ancient pieces of DNA were likely to have been chemically modified in all sorts of ways, making it impossible to resurrect them. So Krings used a molecular trick called PCR* to amplify the DNA, beginning with a short section about one hundred bases long. He found to his delight that while some of the amplifications of this short section resembled modern humans and were therefore probably contaminants, most of them did not.

Were these new bits of DNA really a molecular echo from the distant past? Patiently, Krings used PCR to amplify short sequences

---

* PCR, polymerase chain reaction, allows an experimenter to make huge numbers of copies of a short piece of DNA. It requires that something be known about that piece, however. If the experimenter knows the sequence of two short regions on either side of the piece, he or she can use an enzyme to fill in the intervening region. The reaction is arranged in such a way that, each time this is done, the number of copies of the intervening region doubles. After twenty or thirty cycles there are billions or copies—enough to use standard laboratory techniques to determine the sequence of the intervening region.

that overlapped the first. These pieces could be joined together into a longer fragment, eventually almost four hundred bases long. A variety of control experiments convinced him that this really was thirty-thousand-year-old DNA. To clinch the matter, he sent his material to another laboratory in the United States, where Anne Stone, an expert on Native American mitochondrial DNA, was able to amplify the same sequence.

The sequence of bases in the Neandertal DNA (for such it almost certainly is) shows distinct differences from the equivalent region in modern human DNA. The lineage that led to this western Neandertal really had been separate from us for hundreds of thousands of years, just as the fossil discoveries in Atapuerca had suggested.

More Neandertal DNA sequences will have to be studied—particularly from the Neandertals of Central Europe and the Middle East—if the true place of the Neandertals in human evolution is to be determined. But don't try this at home—Krings very sensibly added a warning at the end of his paper, pointing out that the Neander valley skeleton is particularly well preserved. Most Neandertal bones are unlikely to yield any DNA, unless tests to determine the extent of molecular damage are carried out first. If the drills and saws of molecular paleontologists are not used with caution, irreplaceable Neandertal remains may end up looking like Swiss cheese without telling us anything new about their DNA.

One other intriguing feature of the Neandertal sequence emerged from Krings's study. Although it is indeed very different from our DNA, some mitochondrial DNA sequences from modern humans are substantially different from each other as well. Some sequences taken from African pygmies and Australian aborigines, for example, are about two-thirds as different from each other as human sequences are from the Neandertals. If we knew very little about our own DNA but had just by chance obtained a few sequences from very isolated human groups, then the Neandertal DNA might not seem particularly unusual.

The Neandertals were far, far closer to us than our nearest living relatives, the chimpanzees. Some of them may have been even closer to us than the morphologically extreme member of its race that Krings examined. After all, the bones of that Neandertal were found in the isolated Neander valley, far from Eastern Europe and the Middle East where the real evolutionary action was taking place.

THE FINAL DAYS OF THE NEANDERTALS

In Western Europe, probably between about 35,000 and 28,000 years ago, the ancient western Neandertal lineage came to an end. Their disappearance coincided with the arrival in Southern Europe of the Cro-Magnon people and a number of other groups, who were virtually indistinguishable from modern humans and had far more sophisticated cultures. Most people have assumed that the Cro-Magnons were simply too clever for the Neandertals and were able to drive them swiftly and mercilessly to extinction.

Yet there is more to the story. The Neandertals were hardly pushovers—the first appearance of the Cro-Magnons and the final disappearance of the Neandertals were separated by at least seven thousand years, a period of time substantially longer than the entire history of recorded civilization.

It was presumed until recently, in the absence of evidence, that the Neandertals were savages indeed and knew nothing of the decorative arts or what we might suppose were the niceties of civilization. While some of their remains show signs that they had been buried, the burials seemed to have been nothing more than simple interments. The bodies were neither accompanied by artifacts nor covered with the pigments used by the Cro-Magnons and other later peoples. But excavations of Neandertal remains from a cave in the Crimea show that spring flowers had been buried along with the bodies, so that these people had invented at least some elements of ritual burial. And some of the skeletons show signs of severe injuries that had subsequently healed, indicating that the victims must have been nursed through their trauma by other members of the tribe.

Recent important discoveries suggest that toward the end of their career, the Neandertals might have progressed considerably in their technology, though it is not yet clear whether this happened because of contact with the Cro-Magnons and other more advanced peoples or whether they accomplished these advances without outside help.

In early 1996 Eric Boëda and a group of French and Syrian colleagues working at a remote site in western Syria discovered stone tools, at least forty thousand years old, that had traces of black pitch adhering to them. Careful chemical analysis showed that the pitch must have been heated and melted in a fire. Probably the pitch had been used to bond the tools to wooden handles. Were the makers of these sophisticated tools Neandertals? Boëda thinks so, since he recently found a Neandertal skull in the same deposit.

Until this discovery, the oldest tools showing traces of pitch as an adhesive were a mere ten thousand years old. By an odd coincidence those tools had also been found in the same area, which raises the intriguing possibility of cultural continuity between the Neandertal makers of the older tools and the more modern people who made the more recent tools.

Perhaps the most remarkable discovery of an advanced Neandertal artifact was made in the summer of 1996 by a team headed by Ivan Turk of the Slovenian Academy of Sciences. His team was excavating deposits in a cave in hilly country to the west of the city of Ljubljana, near the current border between Italy and the former Yugoslavia. There were many signs that the cave had been occupied by Neandertals, including some typical stone tools buried in the cave floor. But along with these tools the excavators also found the remains of an astonishingly advanced artifact. It was a piece of the femur of a juvenile cave bear in which four neat circular holes had been punched, all along the same side. It seemed likely to them that this fragment formed part of a flute.

If it was a flute, then it is the oldest ever discovered. It has been dated using electron spin resonance measurements of cave bear teeth that had been found in the same layer; the results show that it was made, and perhaps played, somewhere between 40,000 and 80,000 years ago. It must already have been part of a long musical tradition, for the punched holes are precisely placed and beautifully round (Figure 7–4). Their position suggests that the flute could have been used to play a diatonic scale, with full and half tones.*

This remarkable object implies a maker who had a deep understanding of how to coax a variety of sounds from a hollow tube. The Neandertals might not have made it themselves; they may have obtained it through trade or warfare with its real makers, perhaps a more

* It has recently been suggested that the bone was not part of a flute at all, and that the holes were punched by animal teeth. The edges of the holes show no smoothing of the type that would be expected if the instrument had been used. This flute-versus-natural-object argument is likely to continue, and its final outcome is uncertain. But the holes are remarkably round and regular in their placement. The flute may have been broken in its manufacture before it had a chance to be used.

Figure 7–4. A possible Neandertal flute on the left, and on the right, a closeup of the strikingly round yet puzzlingly unworn holes in the bone.

advanced tribe of early Cro-Magnons living in the vicinity (but for whose existence there is no evidence). But if the Neandertals were as brutish as they are supposed to be, why would they have wanted such an object?

The more we learn about Neandertals, the more like us they become. If they really could use a flute to make music, they must also have invented complex ritual occasions for doing so. Any impartial observer would agree that this constitutes fully human behavior. Indeed, it seems less than kind of us to relegate these flute-playing hominids to a subhuman status—particularly when most of *us* are not able to play flutes!

The western Neandertals, in spite of their isolation, showed remarkable adaptability right up to the end. They seem to have invented their own moderately advanced culture of stone tools and implements. It has been named the Chatelperronian and was quite distinct from the somewhat more complex Aurignacian culture of the Cro-Magnons who replaced them. We will probably never know whether this burst of invention at the end of their career was stimulated by the arrival of the Cro-Magnons, or whether they would have done it on their own.

They may have gone even further than these suggestions imply. Some years ago a collection of tools and decorative objects was found in the caves of Arcy-sur-Cure, near Auxerre, in the Nivernais region south of Paris. The artifacts include shells and other objects with holes drilled in them, so that they could be dangled on thongs. They have been dated with some assurance to 35,000 years ago and are among the very earliest such objects found anywhere.

The Arcy-sur-Cure finds are very typical of objects made by Cro-Magnon people. But were the people who made them, or at least possessed them, really Cro-Magnon? Tantalizing bits of skull and other bones were found along with the artifacts, but these bits were so fragmentary that it seemed at first that they could tell nothing about the people who lived in the caves.

Then in 1994 a group led by Jean-Jacques Hublin of Paris's Musée de l'Homme employed a CAT scanner to look inside some of the skull fragments and visualize the semicircular canals of the middle ear. For some reason that has yet to be discovered, the shapes of these canals in Neandertals and modern humans, including Cro-Magnons, are clearly different. The CAT scans showed unequivocally that the bone fragments were Neandertal, not Cro-Magnon.

It now seems that Neandertals occupied the cave of Arcy-sur-Cure and may have been the owners—if not the makers—of those advanced artifacts. Possible scenarios abound. Did they steal the objects, or acquire them through trade? Did they learn how to make them by imitation? Alternatively, was the cave really occupied by Cro-Magnons, who were in the habit of hunting Neandertals, bringing their heads back, and decorating these gruesome trophies with necklaces and earrings? The most daring possibility of all: Did the Neandertals invent these objects on their own—and then teach the Cro-Magnons how to make them? Whatever the explanation, western Neandertals and Cro-Magnons apparently came into contact, perhaps repeatedly, and some technology could have been transferred.

The fate of the western Neandertals is made more poignant by these new discoveries. Were they driven utterly to extinction? No

DNA sequences resembling that found by Matthias Krings have yet been found among living humans, so there is no direct evidence that the Neandertals left behind any genes when they disappeared just under thirty thousand years ago. But even though their incipient speciation seems to have been aborted, perhaps they did manage to exchange at least a few genes with the Cro-Magnons. If so, then a Neandertal-like sequence may be found at some future time in a modern human, providing direct evidence of such an exchange.

Further, as Erik Trinkaus points out, when some of the Neandertals escaped from their isolation in the west, they seem to have made repeated contact with more culturally advanced humans in middle Europe and the Middle East. If gene exchanges between these two groups did not take place, then it may be our loss.

The Neandertals lived a life of unimaginable rigor, able to survive and even thrive during the great climatic swings of the most recent ice ages. Yet in spite of all these physical challenges, their brains grew larger and their culture became more complex over time, just as ours did. In their lineage, as in ours, cultural evolution did not preclude physical evolution. Acquiring more elaborate culture did not slow down their physical evolution—if anything, it seems to have accelerated it.

Modern humans, too, were pushed in roughly the same direction. Why? Not because we have different genes from the Neandertals, but because when modern humans met Neandertals, both groups had roughly the same set of alleles, though in different proportions. Remember that Neandertals are not much farther from us genetically than the maximum distance between living human groups. Just as with human groups living today, the differences between the Cro-Magnons and the Neandertals must have been primarily allele frequency differences, not the presence of an allele in one group and its absence in the other.

These allele frequency differences were expressed in many ways. Some of them had to do with brain size increases.

The pressures to increase brain size must have been very strong, powerful enough to have found a way around equally strong evolutionary restraints. The shapes of the fronts of the Neandertals' skull were constrained by the requirement for strong jaw muscles, particularly the temporalis muscle that attaches to the side of the skull and passes beneath the cheekbone to attach to the jaw.

You can feel this muscle by touching the area above your temple and clenching and unclenching your jaw. In modern humans it is a feeble thing, but in the Neandertals it was far larger. The projecting browridge above the eye sockets and the sloping forehead, which to us gives the Neandertal skull such a primitive appearance, were needed to provide both room and a substantial attachment point for this muscle. The front part of our own skulls has grown larger, resulting in our near-vertical foreheads and a reduction in the thickness of the temporalis muscle sheet. This option for increasing brain size may not have been possible for the Neandertals, since it would have weakened their jaw muscle too much.

We do not know the diet of the Neandertals or how, if at all, they prepared their food. We do know that present-day Eskimos, who probably live under very similar conditions, use their teeth as tools to make many different items of clothing and artifacts, and it seems likely that the Neandertals did the same. They are, perhaps, a vivid example of the old saw that you are what you eat.

Given these constraints, which prevented their foreheads from becoming more vertical as their brains expanded, how could they have acquired larger brains? The only way, it would seem, was by expanding the skull to the rear. The occipital boss, so pronounced in late western Neandertals, is the result of this final expansion. It had not yet appeared by the time of the people of Sima de los Huesos.

In view of the sometimes substantial variation in skull shape among living human groups, it is not difficult to see how the different shapes of our skulls and those of the Neandertals could have been produced, not by the appearance of new genes, but simply by shifts in the frequencies of preexisting alleles.

LADINS AND NEANDERTALS

The tale of the western Neandertals casts light on our own recent and continuing evolution. Although they lived in a kind of parallel universe, following a different evolutionary path from ours, in the end they seem to have become remarkably like us. Then they either disappeared entirely or managed to contribute some of their genes to our gene pool.

But what do we really mean by "contribute to a gene pool"? The phrase is glibly tossed off by generations of geneticists and anthropologists whenever they talk about mixtures of human groups, but without specifics the term has little meaning. To make it more concrete, we must recall that the blending taking place among different human gene pools may be having a substantial effect on our evolutionary potential. This will, however, only be the case if the various gene pools that are undergoing the blending are substantially different from each other.

When population geneticists have looked at samplings of the genes from different human groups, they have been able to find very few real differences in alleles between groups. Most human groups have roughly the same collection of blood group alleles, enzyme alleles, and so on. The differences lie primarily in the frequencies of these alleles.

Further, the total amount of allelic diversity within a single human group is far greater than the differences between separate groups. According to an estimate made by geneticist Richard Lewontin, variation within groups accounts for at least ninety percent of the total variation in our species.

Sometimes we can see in our own species milder examples of what happened to the Neandertals. Geneticists have recently examined a people who live in the valleys of the Dolomite Mountains in northeast Italy and speak a language called Ladin. The origin of their language is a mystery, yet the people who speak it seem no different in appearance from their neighbors who speak other languages. In addition to speaking their unusual language, some of them carry unusual mitochondrial chromosomes, as genetically distant from those of their neighbors as the mitochondrial chromosomes of Africans. Like the origin of their language, the origin of their mitochondria is a mystery.

Their story illustrates the very different way in which mitochondrial and nuclear chromosomes are inherited. Some of the ancestors of the Ladins may once have been very different in physical appearance from the surrounding tribes, but over time, as they married across tribal boundaries, their nuclear genes shuffled and recombined. The result was that any physical differences between them and their neighbors disappeared. Their mitochondrial DNA, however, did not recombine, and a few copies of it came down a separate lineage as an echo of their distant past. Their language, too, has remained separate, and while it has undoubtedly changed greatly from the ancestral language of the Ladin people, it has retained its unique identity.

The Ladins have contributed to the cultural diversity of the northern Italians and must have added genetic diversity as well. But except for those telltale mitochondrial chromosomes, no sign of their genetic contribution persists.

To find out whether the same process happened to the Neandertals, we need more evidence from their DNA—not just from their mitochondrial chromosomes but from their nuclear genes as well. Will their nuclear genes, as I strongly suspect, turn out to be very like our own? Did they, too, have ABO blood group alleles, and Duffy-a and -b alleles, and perhaps the same set of dopamine receptor alleles, only in a slightly different mix? Because copies of nuclear genes are likely to be very rare in Neandertal bones, finding them is probably beyond the reach of present-day technology. Yet I suspect it will only be a matter of time before the DNA hunters are successful. And when these genes are found, I predict they will show that we are closer to the Neandertals than mitochondrial DNA alone would suggest.

Now let us leave the remarkably parallel story of the Neandertals and turn to the question of how our own species has reached its current state. As we will see in the next chapter, the evolutionary histories of chimpanzees and gorillas have been very different from our own. If chimpanzees from everywhere in their natural range—which extends across the entire width of tropical Africa—were suddenly to be mixed together in a huge involuntary gene-blending experiment of the type that our species is currently undergoing, the effect on their physical appearance and behavior would be slight and perhaps even unnoticeable. This implies that the differences between human populations must somehow be greater than the differences between chimpanzee populations—even though it turns out evidence from mitochondrial chromosomes seems to point in the opposite direction. New genetic advances means that we will soon be able to glimpse just how our evolution took place and why it has been so similar to that of the Neandertals and yet so different from the evolutionary history of the great apes.

EIGHT

◈

Why Are We Such Evolutionary Speed Demons?

The most casual student of animal history is struck by the fact that while most phyletic lines evolve regularly at rates more or less comparable to those of their allies, here and there appear some lines that seem to have evolved with altogether exceptional rapidity.

GEORGE GAYLORD SIMPSON, *Tempo and Mode in Evolution* (1944)

CHIMPANZEES ALL THE WAY BACK

DNA evidence tells us that the evolutionary lineage that led to the gorillas split off from ours perhaps seven to nine million years ago; the lineage that led to the chimpanzees departed along its own path between five and six million years ago. During much of that time the evolutionary stage on which these apes trod was a narrow one, for they have apparently always been confined to the forests of sub-Saharan Africa. Our own ancestors, in contrast, beginning at an unknown point somewhere in Africa, spread first through the entire African continent and later through all of Eurasia.

On the basis of this evidence alone, we might expect that the great apes have evolved less than ourselves. They have not been such brave explorers of new environments. Like the orangutans of Borneo and Sumatra, relatively unchanged for at least two million years, the chimpanzees and gorillas may be stuck in a kind of evolutionary time warp.

Because chimpanzees and gorillas are clearly different from each other, they can hardly have been in complete evolutionary stasis. But, at least over the last few million years, they do not seem to have undergone the enormous changes that can be seen in our own lineage.

I can make this statement because of the remarkably complete fossil record that we have for our species. Although paleontologists are always complaining about how fragmentary it is—and indeed, vast areas of the world and vast spans of time are still empty of traces of our ancestors—the record is now sufficiently good that we can get a fairly clear idea of what happened. Perhaps the most striking finding is that the further back in time we go, the more chimpanzeelike our ancestors become.

This allows us to make a prediction: The evolutionary history of chimpanzees for the last five million years should be, by comparison to our own, supremely uninteresting—chimpanzees all the way back.

The strongest evidence for the chimpanzeelike nature of our own ancestors was found in 1992 by Gen Suwa of the University of Tokyo, along with the Berkeley paleoanthropologist Tim White and an international team of field-workers. As these patient anthropologists were carefully quartering a stretch of eroding ground at a desolate site in the hot, dry Afar region of Ethiopia, Suwa found a single molar tooth peeking out from the clay.

He and the others then began to comb every inch of the surrounding area, soon finding bits of skull and arm bone nearby. The most exciting discovery was made by Alemayehu Asfaw, an Ethiopian team member. It was a fragment of the lower jaw of a young child, with the deciduous teeth still in place. To the excited anthropologists, this fossil seemed clearly different from anything previously found in the area. In particular, it was different from the jaws and teeth of the famous Lucy skeleton that had been discovered, two decades earlier and not very far away, by Donald Johanson. For one thing, these hominids lived about 4.4 million years ago—almost a million years earlier than Lucy.

The teeth are very like those of chimpanzees, more so than those of any of our later ancestors in the fossil record. The relatively thin enamel was much more typical of chimpanzees and other apes than of humans and was only half as thick as Lucy's.

By sheer luck, White's group also found a bit of the base of a skull. It included part of the foramen magnum, the opening through which the spinal cord passes. This very diagnostic feature can tell us a great deal about a hominid's habitual posture. The heads of modern humans are balanced on top of the spinal column, and our foramen magnum is more or less at the center of the base of the skull. But in the ancient hominid from the Afar, the opening was much farther back, toward the rear of the skull, almost as far back as in present-day chimpanzees.

In something of an excess of taxonomic zeal, White and his coworkers named their find *Ardipithecus ramidus*, thus creating for it a new genus as well as a new species. They were following that old tradition among fossil hunters: discoverers of new and exciting fossils have always tended to give them new names, even though it is not really possible to tell from the bits of bone themselves how much their owners differed genetically from similar creatures or even from ourselves.

Since this first discovery, more *Ardipithecus* fragments have been found. While these latest discoveries show that it was probably able to stand upright, its posture was definitely not the same as ours. Standing fully erect would have been as tiring for these hominids as it is for present-day chimpanzees, since their heads projected so far forward. They must have spent only rather short periods fully upright.

In the first paper published about this find, White and his colleagues admitted that one would be hard put to distinguish *Ardipithecus* from a chimpanzee. There were only some slight differences in the teeth, the small shift in position of the foramen magnum, and some minor additional features of the skull.

The resemblance to chimpanzees extends beyond skeletal similarities, to embrace the world in which *Ardipithecus* lived. These hominids, like the majority of present-day chimpanzees, spent their lives in fairly dense forest. The nature of their habitat can be gathered from a variety of bits of evidence. Many animal bones have been discovered in the same slopes as the *Ardipithecus* fragments, including those of numerous colobus monkeys. Today such animals are typical of dense closed-canopy forest, and their ancestors most likely lived in a similar environment. It was a setting quite different from the open and more fragmented woodlands inhabited by the later hominids in our own lineage, such as Lucy.

These findings mesh with what we know about *Ardipithecus*, particularly the thinness of the enamel on their teeth. In the forest their diet would have been rich in fruit and tender leaves. Only later, when their descendants had moved out into more open country, did their diet change to include more difficult-to-chew foodstuffs. Lucy's thicker tooth enamel was the evolutionary result of that change.

Ardipithecus is the most chimpanzeelike of our forebears to have been discovered, and it also happens to be the oldest. But the overall resemblance to chimpanzees persisted for a long time in our lineage. In 1924, when Raymond Dart found the first skull of our strikingly ape-like ancestors in material excavated from a South African quarry, he named it *Australopithecus*, "southern ape." The date of this young child's skull is still a bit uncertain, but it is probably about two million years old—less than half as ancient as *Ardipithecus*. It too, however, is so distinctly chimpanzeelike that the British scientific establishment initially dismissed it as an ancestor not of humans but of apes.

A few years ago, at the South African site of Sterkfontein, paleoanthropologist Ron Clarke demonstrated the similarities to me. Side by side, on a rough wooden table, he placed a cast of a 2.5 million-year-old *Australopithecus* skull and a skull from a modern chimpanzee. I was struck, just as those establishment authorities of the 1920s had been, far more by their overall resemblance than by their differences, which were primarily in the teeth. And this resemblence had persisted even though the *Australopithecus* skull had undergone two additional million years of evolutionary change since the time of *Ardipithecus*.

Of course, those remote ancestors of ours must have been different from chimpanzees. Chimpanzees, too, have undergone five or so million years of separate evolution, and their remote ancestors must have been different from the chimpanzees that we see today. But the differences must have been relatively minor.

Paleoanthropologists have probably not yet found fossils that date all the way back to the human-chimpanzee branch point. Through a geological accident, fossil-rich deposits that date to the period just before *Ardipithecus* have been hard to find in East Africa. But some are being discovered, and as paleoanthropologists scan these deposits they will continue to close the gap between our ancestors and those of the chimpanzees. One day soon a lucky fossil-hunter will find pieces of a fossil that is very close to the common ancestor. The bones will probably be discovered projecting inconspicuously from a baked and eroding hillside somewhere in Ethiopia, but they might turn up anywhere in Africa—even in West Africa, which until some recent discoveries was thought to be empty of hominid fossils.

It is safe to predict that these ancestors will turn out to have been even more apelike than *Ardipithecus* and to have had chimpanzee-size brains, chimpanzeelike dentition, and many other chimpanzeelike features. They were probably able to walk upright only for short periods and lived very like present-day chimpanzees, banding together into small groups and roaming a fairly dense forest. They probably lived five million or more years ago. If so, this date will match extremely well with dates that have been inferred from DNA and from protein molecules.

Such ancestors, when they are discovered, will also be the first true ancestors of chimpanzees to be discovered in the fossil record, despite the better part of a century of African fossil-hunting. This dearth is not for lack of looking—any bone that a paleontologist finds that might have belonged to an ape or a monkey will receive just as much scrutiny as one that might have belonged to our own ancestors.

The usual explanation given for the lack of fossil remains of chimpanzees or gorillas is that these apes have always lived in tropical forests. As soon as they died, the parts of their bodies that were not eaten by scavengers must have been destroyed by the plentiful insects and bacteria on the forest floor. This argument, however, has recently lost much of its force. Even though *Ardipithecus*, our own oldest ancestor, lived in quite dense forest, a substantial number of fossils of this very chimpanzeelike creature were preserved.

Further, floods or volcanic eruptions occasionally killed our own ancestors and covered their bodies before they could decay. Such accidents appear to have been responsible for a number of famous fossil finds, such as the First Family, a large collection of three-million-year-old Australopithecine bones discovered by Donald Johanson in Ethiopia. The ancestors of chimpanzees and gorillas lived in volcanic areas for just as many millions of years as our own ancestors, and it seems odd that there were no opportunities for such scientifically happy (though personally distressing) accidents to have happened to them as well.

Other kinds of fossils giving clues to our ancestors, traces that are much more ephemeral than bones, have also been preserved. In 1976 paleontologists in a group led by Mary Leakey were working at Laetoli, a site not far from the famous Olduvai Gorge, when they made one of the most important discoveries in all of paleoanthropology. Hardened into the rock, and barely visible unless the light slanted in just the right direction, were sets of footprints from a variety of animals. Excavation revealed that some of these footprints belonged to two, or possibly three, Australopithecines.

As these hominids trudged across a field of ash that had been freshly deposited by the eruption of a nearby volcano, they left their footprints behind. Then, through a further lucky coincidence, rain fell briefly, just enough to harden the ash but not enough to destroy the fresh prints. This astonishing find provided proof positive that by 3.5 million years ago Australopithecines were already able to walk upright.

Why did such a lucky accident never befall the footprints of chimpanzee or gorilla ancestors? The most likely explanation is that their ancestors were never as numerous or as widespread as the human ancestors. Although fossils of the ancestors of these great apes or of their footprints must be somewhere out in the endless baking gullies of East Africa, perhaps exposed on the surface at this very moment, they are so rare that no one has yet stumbled on them. Whatever first great adventure took *Ardipithecus* on the new path that eventually led to us, the apes had no part in it.

A QUICK OVERVIEW OF HUMAN PREHISTORY

Our own pell-mell evolutionary story has been very different from those of the chimpanzees and gorillas. Many different evolutionary changes, at many different times, have happened that resulted in our big-brained, clever-handed, highly social, language-speaking selves. The ability to stand upright seems to date back at least four million years.* Hands that, like our own, had extreme touch sensitivity and

---

* Upright posture may not be unique to our own lineage. An ape that lived ten million years ago on Sardinia, *Oreopithecus bambolii*, seems to have acquired similar capabilities, perhaps independently.

great manipulative skills evolved 2.5 million years ago. Their appearance coincides with the first stone tools in the fossil record. Substantial increases in brain size and in stature took place about two million years ago. Fire was probably first used at around that time.

The appearance of language is more problematical, since it has proved impossible to trace the ancestral tongues of present-day languages back more than ten thousand years, and even these efforts have been dogged by controversy. But our brains are so hard-wired for the acquisition of language that this skill must have had its genesis a long time ago. It is unimaginable that the spear-throwing people who lived in Germany 400,000 years ago were incapable of communicating verbally with each other during the course of their hunts.

Although writers on the subject tend to point to one or another of these events, particularly the acquisition of language, as the watershed event in the appearance of modern humans, they are all part of a remarkably seamless movement toward creatures like ourselves. We cannot single out one event that catapulted us into "full" humanity.

Further, we have only a few glimpses of our ancestors' diversity. If we knew how genetically variable they were, we could say something about their evolutionary potential, since diversity is a powerful resource for continuing evolution. But because very few fossils can be assigned to a given time period, we do not know the full range of types of people who lived at various points. The people of the Sima de los Huesos are an exception, and even they represent only one tribe. They had a great deal of variation in brain size, while at the same time their jaws and teeth were remarkably uniform. Such glimpses of the variation among our ancestors are unusual. Yet it is just this kind of variation that has powered our rapid evolution.

It is certain that there was enormous diversity among the hominids of about two million years ago, although we do not know how much there was among the subset who were our direct ancestors. Discoveries in East Africa, particularly by Louis and Mary Leakey and their son Richard, have showed that around that time an extremely varied collection of hominids roamed the plains and patchy forests. Small-brained Australopithecines were present in large numbers; they had, by this time, evolved beyond Lucy and the older and more chimpanzeelike *Ardipithecus*. They stood more nearly upright, and their hands were very similar to ours. Randall Susman of the State

University of New York at Stony Brook, who has examined many of these hand bones, strongly suspects that their owners were capable of using tools.

By two million years ago the Australopithecines had split into at least two lineages, one of which was robust and had huge teeth and jaws, and another that was relatively more light-boned or gracile (though compared with us fragile creatures, these gracile Australopithecines were pretty robust as well). (See Figure 8–1.) It is not easy to determine what place any of these Australopithecines have occupied in our own ancestry, though the light-boned ones are generally agreed to have been more like us. To make things more confusing, the distinction between the robust and the gracile Australopithecines has blurred a bit in recent years—fossils showing features that are intermediate between them have turned up in South Africa.

It was at about this time that the much larger-brained *Homo habilis* appeared on the scene. They are the first creatures to be sufficiently like ourselves that general scientific consensus has permitted them to join the genus *Homo*. All anthropologists agree that a major evolutionary step was necessary for them to do so, although it is not clear just how quickly that step took place, or whether it happened all at

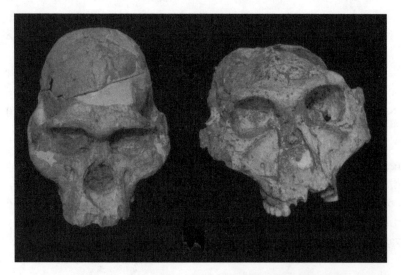

Figure 8–1. Skulls of a gracile Australopithecine ("Mrs. Ples," Sts5) on the left, and of a far more massive robust Australopithecine (SK48) on the right.

once or in piecemeal fashion. An increase in brain size was certainly an important part of this change, but not the only one. *Homo habilis,* like the Australopithecines, was short in stature; an increase in body size to something more like that of modern humans seems to have come much later than these first increases in brain size.

When paleoanthropologists penetrate this far back into the past, they simply do not have enough information to follow the stories in detail. As nearly as we can tell, over a span of perhaps half a million years the brain sizes of the hominids now grouped together as *H. habilis* increased by about fifty percent over those of the Australopithecines. This increase is certainly striking, but it must be remembered that it is based on a mere six skulls—the only ones that are sufficiently complete that the sizes of the brain cavity can be measured.

The most famous of these skulls, known as KNM-ER 1470, was found by Bernard Ngeneo, an associate of Richard Leakey, in 1972. It is relatively well preserved and is shown on the left of Figure 8–2. Its owner lived almost two million years ago, in quite open country near a lakeshore in what is now northern Kenya. The skull has space for a very large brain, with a volume of perhaps 750 cubic centimeters. Here *large* is a relative term, since its volume was equal only to two cans of soda, a little more than half the volume of our own brains. It was, nonetheless, more than half again as large as the brain of an average Australopithecine.

Had the braincase been destroyed while the rest of the skull remained intact, this find would almost certainly have been classified as an Australopithecine. Like an Australopithecine's, the face is very prognathous or forward-thrusting and very broad. KNM-ER 1470 seems to have been classified as *Homo habilis* primarily by virtue of its mighty brain.

Some of the other *Homo habilis* skulls are very different in size from KNM-ER 1470. Although they vary greatly in their completeness and their overall appearance, it has nonetheless been possible to make an approximate estimate of their brain sizes. While some were large, others were much smaller and actually appear to have overlapped those of the Australopithecines.

So if their brains were so small, why have these creatures been grouped with *Homo* rather than *Australopithecus?* One important reason is that their faces tend to be narrower and more vertical—less

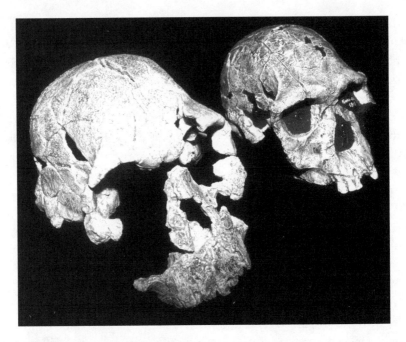

Figure 8–2. The larger, more fragmented, but clearly more robust KNM-ER 1470 on the left, and the smaller KNM-ER 1813 with its narrower and more vertical face on the right.

prognathous or forward-thrusting—than the more chimplike faces of the Australopithecines. (See the skull named KNM-ER 1813 in Figure 8–2.) The evolution of a relatively narrow and vertical face, which we think of as typically human, is surprisingly separable from the evolution of increasing brain size. Anthropologists appear to be quite accommodating about the characteristics that a hominid need display in order to be admitted to the select company of earliest *Homo*. Either a large brain or a vertical face will do.

The skulls of *H. habilis* pose difficult problems for the scientists who have to sort them out. One of them, known as KNM-ER 1805, has such an extravagant mix of Australopithecine and more human characteristics that it seems to be a blend of the two. The existence of this skull raises an important question: Were there, as most authorities on the subject have concluded, really several different early species of *Homo*? Or could they all have been one species, made up of individuals that varied greatly among themselves, so that KNM-ER 1805 is an example of this mixture of characteristics? What was the real relationship of this mix of hominids to the Australopithecines that some of them resemble so closely?

While this problem may seem highly technical, it actually has a direct bearing on our own recent history, since similar questions have been raised about the evolutionary events that lead directly to *Homo sapiens*. If multiple species of *Homo* existed two million years ago, then all but one of them must have died out, since by definition the various species would have been unable to exchange genes once they had parted genetic company. But did they? What really happened? Did ecological upheavals provide multiple opportunities for gene exchange, of the sort now happening between the Bornean and Sumatran orangutans and that possibly took place between Neandertals and modern humans?

THE BEGINNINGS OF LANGUAGE

The brains of *H. habilis* have long since disappeared, but it is possible to learn something about them from the impressions they left behind on the interiors of the skulls. Careful study by Ralph Holloway, Phillip Tobias, and others has shown that the brains of the Australopithecines really do seem to have differed from those of the best-preserved and largest-brained *H. habilis*.

The differences were more profound than mere size. They were concentrated in the front and side regions of the brain, which in modern humans are involved with the production and understanding of speech. These regions appear to have undergone a great enlargement in *H. habilis* compared with the Australopithecines. So, apparently, did the mysterious cerebellum or "little brain," nestled near the brain stem. Recently, functional magnetic resonance imaging (MRI) scans on volunteers have shown that this primitive brain structure is very active during speech. Until the advent of MRI, it had been thought that the cerebellum was concerned only with fundamental abilities like balance, but it too has taken part in our rapid recent evolution and has acquired important new functions.

Did all these changes in its brain mean that *H. habilis* could talk? Probably nowhere near as readily or as incessantly as we do, but it could doubtless use its brain in ways that Australopithecines could not. Though its language skills were probably extremely limited, *H. habilis* was probably capable, to some degree, of applying labels to things and processes in the world around it, and of communicating these labels to other members of its tribe.

There is likely to have been another difference between *H. habilis* and its ancestors. When researchers attempt to teach chimpanzees sign language, or the use of symbols as communication tools, it takes endless persuasion and reinforcement to produce results that are rather marginal and difficult to interpret. One of the most obvious hallmarks of our own species is the way in which young children acquire language rapidly, in an eager, unforced, and enthusiastic way. If *H. habilis* were able to speak in some fashion, as the shape of its brain seems to indicate, then the joy of communication must have been evolving at the same time as that capability. This enormous step on the road to humanity, like all the others, was a complicated one and did not happen instantaneously.

PEOPLE MORE LIKE US

This plethora of African hominids was eventually supplanted by a more advanced, and to all appearances even more successful, hominid known as *Homo erectus*. The new hominid was almost certainly a direct ancestor of ours. Its earliest fossil remains date from almost two million years ago in both Africa and Asia, but it probably made its actual debut about half a million years earlier. *H. erectus* was, so far as we know, our first ancestor to leave Africa and travel through the Old World.

The earliest complete African skeleton of this hominid happens to have been superbly well preserved. Australopithecines and *H. habilis* tended to be small in stature, but this adolescent boy was very tall and probably would have reached six feet at maturity. His brain, too, was larger than that of even the brainiest *H. habilis*, approaching nine hundred cubic centimeters for the first time.

The triumph of *Homo erectus*, while it seems eventually to have been complete, was nonetheless anything but instantaneous. The gracile or light-boned Australopithecines did vanish at around the time that *H. erectus* appeared, an event that has haunting resemblances to the much later replacement of Neandertals by Cro-Magnons. This earlier replacement process, however, took as long as half a million years, not a mere seven thousand.

The robust Australopithecines managed to persist in various parts of southern and eastern Africa until less than a million years ago. Because these animals shared much of the eastern region of Africa with *H. erectus* for a million years or more, this suggests that the two hominids had such divergent ways of life that they were able to coexist without coming into mortal conflict.

Because no Australopithecine fossils have been found outside Africa, we are fairly confident that it was *H. erectus*, not the Australopithecines, who first migrated through the Middle East and into Asia about 2.5 million years ago. At that time the Pleistocene ice ages still lay in the future and the climate was relatively mild, so that no climatic barrier stopped them from venturing into Europe or further into Asia and probably back again. Even though we have found only Australopithecines and *H. habilis* from that time in Africa, early representatives of *H. erectus* were probably there too.

The first indication of these early population movements comes from some remarkable new discoveries of ancient stone tools in Israel by Avraham Ronen and his colleagues at Haifa University, and in Pakistan by Robin Dennell of the University of Sheffield. In both cases the tools have been dated quite firmly at about 2.5 million years ago— a remarkable finding which makes them as old as any that have been found in Africa. As soon as our *H. erectus* ancestors invented such tools, it seems, they used them to conquer new worlds. It has generally been assumed that stone tools were first used in Africa, but these new finds raise the possibility that they were invented somewhere else and taken back to Africa by *H. erectus* as they moved back and forth over a wide swath of territory.

The regions into which *H. erectus* ventured were very different climatically from the deserts and exhausted badlands that make up so much of today's Middle East and Southwest Asia. Sometimes one can obtain a vivid glimpse of that ancient world.

In 1994 a group from the Georgian Academy of Sciences began to excavate some old building foundations in the city of Dmanisi, near Tbilisi in Georgia. In the floors of these ancient basements they found deep pits that had been dug by the medieval inhabitants for grain storage. After the debris that filled these pits had been carefully removed, the scientists found that the pits themselves could be used as probes into far deeper strata. The walls of the pits first yielded ice age animal bones. Then, some distance below them, the archaeologists discovered a hominid jawbone, some stone artifacts, and some earlier animal bones that had been missed or ignored by the original medieval diggers.

The jaw shows clear similarities—notably, its lack of a well-developed chin—to *H. erectus*. It has now been dated to between 1.6 and 1.8 million years ago. The animal bones are evidence that the Georgia of those days was a very different place, for among them are bones of giraffes and other warm-weather animals. The mild climatic conditions meant that the animals that are now confined to the East African savannas were far more widespread in those days.

The Israeli site of 'Ubeidiyah, in the Jordan valley some four kilometers south of the Sea of Galilee, also lay in this salubrious region. Excavations starting in 1960 have turned up a large number of remains of animals dated to about 1.4 million years ago. Here too there were giraffes, elephants, and antelope, showing that the climate and range of fauna were very similar to those of East Africa today. Early humans lived in the Jordan valley as well—the archaeologists have discovered primitive stone tools and some very fragmentary bones that almost certainly belonged to *H. erectus*.

Surely, one might think, the more adventurous among those bands of *H. erectus* would soon have taken the path leading through such a pleasant valley directly into Europe. It seems unlikely that they dithered at 'Ubeidiyah for four hundred millennia or so before finally venturing west. More probably we have simply not yet found the earliest traces of hominids in Europe itself. Some of those yet-to-be-discovered early migrations eventually gave rise to the people of the Gran Dolina and the tribe whose bones were preserved in the Sima de los Huesos.

PREDECESSORS OF MARCO POLO

We are not sure where and when people closer to us than *H. erectus* first appeared, but some quite modern-looking skulls, dating from about 100,000 years ago, have been discovered at a variety of sites in southern and eastern Africa and in the Middle East. They have been given a variety of names and are generally referred to as archaic or early modern *Homo sapiens*. These people, with their large brains, quite vertical faces, and high foreheads, were not, however, quite like us.

They were more robust, their teeth were larger, and they had primitive-looking characteristics like pronounced browridges. It appears that even our immediate ancestors, close though they were to us, continued to evolve over the last 100,000 years.

Moreover, these brainy ancestors of ours were not instantaneously all-conquering. In the Middle East they lived in the same region and at roughly the same time as other groups who had more Neandertal characteristics—and they may, though the evidence is only suggestive, have interbred with them.

It took the Cro-Magnons seven thousand years to replace the western Neandertals. Now evidence indicates that, at the other end of Eurasia, our ancestors shared the world of *H. erectus* for an even longer period. The overlap of *H. erectus* with people like us did not last for half a million years, like the overlap of *Homo* and the gracile Australopithecines, but it did last for much longer than anybody had imagined until recently. Evidence for this astonishing persistence of *H. erectus*, in spite of what must have been innumerable contacts with our own species, was found recently on the island of Java.

Of all the Indonesian islands, Java is the most heavily populated, and its landscape has been greatly modified by human activity. The countryside is now cultivated so intensively that terraces have been built right to the tops of quite high mountains. The farmers who work them must spend most of each day climbing up to them and back. Despite such efforts the island is unable to feed its burgeoning population and is a net rice importer.

During its less crowded past, however, Java was a rich prize. Here Buddhist, Hindu, and Mogul empires rose, collided, and fell. They left behind the remains of hundreds of temples, particularly in the central highlands. Earthquakes and volcanic eruptions have shaken most of these ancient structures into rubble and covered them with layers of ash. Only a tiny fraction of them have recently been excavated and rebuilt.

Even though the oldest temples are relatively recent, dating to about the seventh century A.D., their builders are unknown. Our ignorance of Java's more ancient history is even more profound. Were we able to read it, however, that history would surely be long, eventful, and highly informative about human evolution.

During the last 1.64 million years, whenever the world went through a period of glaciation and the sea level dropped, Java became part of a much larger landmass. This huge peninsula extended down from Southeast Asia, encompassing most of the islands of Indonesia and the shallow Sunda continental shelf on which they rest. There were often times during the last part of this period when modern humans were able to migrate south across this shelf from what is now Indochina.

Some of these waves of peoples penetrated as far as New Guinea and Australia, and there is growing evidence that they first arrived in those regions a long time ago. The human history of Australia, in particular, may be older than scientists had supposed until recently. Starting quite suddenly, about 140,000 years ago, great fires swept again and again through huge areas of that continent. They transformed the landscape, destroying dense acacia forest and replacing it with open grasslands and scattered eucalyptus groves. There is a strong possibility that these fires were set by humans. There is far less controversy over other and firmer evidence that people arrived in northern Australia at least sixty thousand years ago.

The migrants must have passed over the Sunda landmass, including Java, on their way to Australia and points east, yet on Java itself there are no fossils or artifacts that can be traced to migrants from this period. The earliest remains of modern humans found on the island date to a mere ten thousand years ago, although a modern skull that might be fifty thousand years old has been discovered on Borneo to the north.

The only Javanese fossil bones that predate this relatively recent find are not modern at all. And rather than a record of migration and change, the story they tell appears to be one of virtual stasis.

These ancient peoples lived on or near the Solo River, which springs from the central Sangiran Plateau, flows muddily across most of the width of the central part of the island, and eventually wends its sluggish way to the north coast. While the river has been a feature of the landscape for a long time, it has often changed its course, carving out a wide floodplain in the process. Here and there on this plain, the floods that shaped it have left traces, in the form of the bones of drowned animals and hominids.

The remains tended to collect at bends where the onrushing flow of the river slowed briefly, allowing them to settle to the muddy bottom and be buried. After the river shifted its bed, these old accumulation sites were sometimes left high and dry. It was detritus from these ancient disasters that in 1891 provided Eugène Dubois, a Dutch doctor, with the first clues to these ancient people.

Dubois put the hominid fossils that he discovered, chiefly crania, into the genus *Pithecanthropus*, because they seemed extremely primitive and ancient—a true missing link between apes and men. It now seems quite certain that these hominids were not unique but were really representatives of *H. erectus.*

Subsequent excavations by the Dutch occupiers of Java turned up even more of these fossils, including one enormous treasure trove of hominid and animal bones found near the village of Ngandong in 1931. Unfortunately, except for a few skullcaps, this entire collection of 25,000 bones, including who knows what treasures, has now disappeared.

The dating of these Javanese specimens, however, has turned out to be extraordinarily difficult. As we saw, nothing—not even massive stone temples—lasts long on Java. Its landscape has been continually rearranged by volcanic eruptions, floods, and earthquakes, which makes its geology much harder to understand than the more stable geology of East Africa. Moreover, the exact provenance of many of the fossils is often lost, because farmers who find bones eroding out of the banks of soft earth immediately pull them out of their geological context and sell them.

This unfortunate process is continuing today. A sad little museum at Sangiran, not far from the banks of the river, enshrines a few extremely sketchy exhibits—not, luckily, including any of the precious crania. Every day, at the base of the steps leading to the museum, jostling hawkers try to sell bits of monkey skull and suspiciously symmetrical stone tools to a scattering of gullible tourists. The real fossils, I was told, are sold to private collectors and disappear.

Carl Swisher, from the Geochronology Center in Berkeley, California, is a pioneer in using the extraordinarily precise argon-argon method of isotopic dating. He has been able to obtain approximate dates for the skulls using individual microscopic crystals of mineral obtained from strata that were probably laid down at the same time. Over the last several years, employing this and other methods, he has obtained a dramatic set of dates that have utterly changed our concept of the history of *H. erectus* on Java.

His oldest date, announced in 1995, comes from strata associated with a juvenile skull that had been found near the town of Mojokerto, in a river valley to the east of the Solo River. Astonishingly, the strata and hence the skull could be as old as 1.8 million years. This date would make it quite as old as the very oldest *H. erectus* yet found in East Africa.

Did tribes of *H. erectus* shake the African dust from their cracked and calloused feet almost as soon as they evolved and sprint the entire length of Asia, ending up in Java? Unlikely—if Swisher's date holds up, it will simply be another piece of evidence that each type of hominid probably evolved long before the time of the earliest fossil of that type that we have yet been lucky enough to discover. The migrations from Africa to Java (and perhaps vice versa) were probably more leisurely.

Yet Swisher's remarkable date increases the mystery surrounding the origin of *H. erectus*. Where—and when—did these people first appear? Perhaps the most intriguing aspect of Swisher's find is that it raises the possibility that *H. erectus* might have appeared first in Asia, more than two million years ago, possibly from some peripatetic group of *H. habilis*, then migrated to Africa!

A year after this discovery, Swisher and his colleagues announced a new set of possible dates, from a different fossil collection, that illuminated the very end of Java Man's long and strangely static history on the island. The dates were obtained from the Ngandong collection of skullcaps excavated by the Dutch in 1931, the remnants of that huge lost treasure trove of hominid and animal fossils. The skullcaps were a great puzzle—over the years they had been assigned different dates by different workers, some of them unnervingly recent.

The Ngandong material is indeed too young to be dated by the argon-argon method, but luckily Swisher could use other techniques that allow bones to be dated directly. The Javanese scientists in charge of the fossils were understandably reluctant to allow him to use bits of the skullcaps themselves, but he did manage to obtain some animal teeth from (he hoped) the same strata as the original collection of bones.

The dates he determined from the animal teeth were indeed very recent—somewhere between 27,000 and 53,000 years old. If these numbers are correct, they show that *H. erectus* managed to survive on Java for almost two million years, finally going extinct at a time so recent that it was only a geological instant away from the present.

Remarkably, Swisher's date places the *H. erectus* extinction at roughly the same time as the disappearance of the western Neandertals at the other end of Eurasia.

These numbers were greeted with disbelief by much of the rest of the archaeological community. Swisher's dates imply that if the first modern humans really did come through Java on their way east 140,000 or more years ago, they must have been passing through *H. erectus* country for well over 100,000 years. It seems astonishing that these two types of people could have ignored each other for so long.

Assuming that Swisher's very old and very young dates are correct, what kind of history can we reconstruct for the *H. erectus* of Java? Were the people of the Solo River somehow sequestered from all the other currents of human evolution? Did they, like the orangutans that also lived on Java during that time, live in jungles so remote that the more modern humans migrating through the area seldom interacted with them? If so, then they would have had to conceal themselves for tens of thousands of years. This behavior would seem to be remarkably timid for hominids whose ancestors had ventured across the whole width of Asia almost two million years earlier. And, if they were so rare and timid, how did they manage to be so abundant in the Javanese fossil record, where no sign of modern humans appears until ten thousand years ago?

Perhaps our assumption of numerous migration routes through Java is incorrect, and the island—however lush, tempting, and richly forested—was really a backwater where *H. erectus* could preserve its primitive way of life undisturbed—rather as we currently suppose all of Europe must have been for the Neandertals.

Another possibility is that, once *H. erectus* reached the verdant forests of Java, they took up a simple and self-sustaining forest life and remained invisible to later modern human migrants. We know nothing about that mode of life, for we know far less about Javanese *H. erectus* than we do about the Neandertals of Europe and the Middle East. Few stone tools have been found in Java. We do not know how advanced (or primitive) the culture of the last *H. erectus* who lived at Ngandong might have been, or how much it had changed during the previous two million or so years. All that information was washed away by the flooding Solo River.

Very recent evidence suggests that *H. erectus* was both more advanced and more peripatetic than has been previously thought. To the east of Java, a string of islands stretches away toward New Guinea. Dominated by towering volcanoes, they are now fairly dry but must have been far lusher in the past. Two of them, Komodo and Rinca, are still home to the gigantic Komodo dragon, a lizard once found throughout the chain but which disappeared from the larger islands the better part of a million years ago.

Because these islands have always been fragmented by narrow ocean channels, even during the height of glaciation, it has been assumed that modern humans were the only ones smart enough to be able to raft across from one island to the other. Yet in the 1960s Theodor Verhoeven, a Dutch missionary, reported the discovery of stone tools on the island of Flores that were about 750,000 years old. His claim was ignored, but it has now been substantiated through careful redating by a group of Australian scientists—in fact, the tools now appear to be almost a million years old. *H. erectus*, it seems, was not simply sitting around in the deep forests while dramatic evolutionary events were going on elsewhere. These hominids may even have been responsible for the disappearance of the Komodo dragons, along with giant tortoises and small relatives of elephants, that lived on Flores at the time.

We can tell one more thing from these fossils. During this huge span of time the Javanese *H. erectus* were not entirely static in their physical attributes: there was a small but significant increase in brain size from the earliest fossils of Trinil and Mojokerto to the last of them at Ngandong. Indeed, the last survivors of *H. erectus* have brain volumes of as much as 1,250 cubic centimeters, which quite comfortably overlaps the low end of brain sizes in modern humans.

Further, they may not have spent all their time hiding in the jungle. One tantalizing hint that some of them were still wanderers is the discovery of ten-thousand-year-old skulls in Australia that have characteristics very like those of *H. erectus*, even though their brain sizes are larger. These skulls, oddly, are more recent than some skulls of more modern humans found in other Australian excavations. This has suggested, to anthropologist Alan Thorne and others, that several groups with clearly different physical appearances colonized Australia over a long span of time. Somehow these tribes retained their distinctive characteristics until remarkably recently—just as *H. erectus* seems to have done on Java.*

The reader bewildered by all of this is not alone. These new results have overturned most of our assumptions about the prehistory of Southeast Asia, raising far more questions than they have answered.

It does seem safe to predict, in the light of our growing knowledge of the Neandertals, that Javanese *H. erectus* were far more advanced and had a far more complicated culture than the apparent absence of artifacts would suggest. The banks of the Solo River, and the eroding gullies and slopes of the Sangiran Plateau from which it flows, are certainly filled with undiscovered deposits. Modern excavation techniques should be able to tell us far more about these people. It is essential that the farmers of Sangiran be educated to leave bones and tools where they discover them. Perhaps this can be done by a system of rewards. And in order to make even more sense out of this story, other parts of Indonesia must also be explored. No traces of fossil hominids have been found on the vast island of Sumatra to the north of Java, for example, but they must be there somewhere.

Did the rich evolutionary ferment that led to our species bubble as briskly on the islands of Southeast Asia as it did in East Africa and Europe? There seems every reason to suppose it did. Indeed, the Neandertal story sketched in Chapter 7 is not unique. Consider the pattern. In Africa *H. habilis* coexisted with and eventually replaced the gracile Australopithecines. Later *H. erectus* coexisted with and eventually replaced *H. habilis*. In Java modern humans coexisted with and eventually replaced *H. erectus*. During each of these long periods of coexistence, gene exchanges between the different groups may have helped to power further evolutionary transformations. While it is likely that we differ from *H. erectus* by more unique alleles than we differ from Neandertals, the great majority of the differences between our gene pool and theirs were probably still differences in allele frequency rather than in alleles. These groups and our ancestors shared a diverse, complex gene pool with immense capabilities for further evolution.

---

* This idea, too, has generated much controversy. The skulls of these Australians might simply have been deformed because their heads had been bound as babies.

THE DIORAMA EFFECT

The fossil record offers only a glimpse of this evolutionary ferment. Between roughly 2.5 million and two million years ago, the savannas and gallery forests of East Africa were the scene of particularly dramatic evolutionary change in our lineage. All during those five thousand centuries, a heterogeneous collection of *H. habilis*, probably along with early representatives of *H. erectus*, were sharing a savage and dangerous world with an equally heterogeneous collection of Australopithecines.

We do not know whether these various hominids lived cheek by jowl or were separated by long distances, though fossils of the Australopithecines have been found over a wide geographic area. But all these groups were surely split into many tribes, perhaps hundreds or even thousands of them. Those tribes must have interacted with each other in innumerable ways, ranging from peaceful coexistence through cautious trade to vicious all-out war. If *H. habilis* and the various Australopithecines really were separate species, the beginnings of this separation may have involved such fierce tribal interactions. Many other nascent separations must have been aborted before they went so far: two separate tribes must often have joined up again, and one tribe or alliance of tribes must often have wiped out another.

To get some idea of the sheer scale of all this, it is reasonable to compare it to something closer to our own time. Consider the recorded tribal histories of the Celtic peoples of Wales, Scotland, and Ireland and of the Saxons of England during the Middle Ages. That period was marked by innumerable fierce external and internal wars, triggered by endless battles over succession to leadership within the various groups and marked by an ever-changing mosaic of alliances and enmities among the rulers who managed to survive. More recently the hatreds of these groups have lessened somewhat, and their gene pools have even begun to blend, but under different circumstances their fragmentation might have increased instead. Extremist groups among them—Sinn Fein, Orangemen, Welsh and Scottish nationalists—are still determined to help accelerate such incipient speciation.

Now imagine such a bloody history lasting not five hundred years but five hundred thousand, and played out over an entire continent. Such an imaginative leap gives us a dim idea of the selective pressures to which our ancestors must have been subjected during the critical half million years of genus *Homo's* emergence. What an immense amount of history must have taken place!

We are forced to imagine virtually all of this history, since the only bits of evidence we have are a few fairly complete skulls, a bucketful of assorted bone fragments, and some tantalizing collections of stone tools. But just because we have no direct evidence for it does not mean it did not happen.

The bones themselves hint that things could be rather unpleasant. Recently, prehistorians have begun to revise their assessment about the role of cannibalism, or at least of savage burnings and hackings, among our hominid ancestors. While it was politically correct during the 1970s to presume that savages were peaceful and noble, the discovery of what appear to be burned, split, and butchered human bones at many places in both the Old and New Worlds has changed the assumptions. (The cuts found on the Gran Dolina bone fragments from the hills near Burgos are just one example.) Such unspeakable practices might have played a much larger role in our past than we have hitherto dared to admit.

I suspect we have difficulty admitting to such things because of what I like to call the Diorama Effect. All those museum dioramas that show happy hominids gathered around their campfires, chipping intently away at stone tools, have lulled us into a sense of the peacefulness of the past. Even scientists fall into the trap of supposing that the past was virtually static, that the various hominids of Africa led separate, essentially event-free lives over thousands of generations, slowly evolving or not evolving as the case may be.

This Diorama Effect has to some degree brainwashed all of us, reinforcing our disinclination to admit just how brawling, ugly, dangerous, and complex were the realities that faced our distant ancestors. Here scientists must defer to novelists, who are well aware of the savageries buried in even the most apparently peaceful breasts. Some of them, like Björn Kurtén, have done an excellent job of counteracting the Diorama Effect.

Most of the events that have driven our evolution have been lost to history. But powerful forces are still influencing us, ever more strongly. It took half a million years for the genus *Homo* to evolve from the Australopithecines. Can we doubt that half a million years from now our species, if it survives, will have undergone even more dramatic changes?

NINE

◈

Bottlenecks and Selective Sweeps

When Adam delved and Eve span,
Who was then the gentle man?

JOHN BALL, inciting the Peasants' Revolt (1381)

The dizzying history sketched so briefly in Chapter 8 marks us out as evolutionary speed demons par excellence. Although we have concentrated on the fossil record, traces of all this speed have been left, not only in our bones, but also in our molecules. Recently my colleagues and I have begun to look more intensively than ever before at human DNA molecules and those of our nearest primate relatives, to find out why we differ so dramatically from them. The most vivid of these molecular stories has emerged from work with our nearest relatives, the chimpanzees.

Pascal Gagneux has spent much of his scientific life investigating chimpanzees, both in the wild and in the laboratory. Deep in the forests of the Côte d'Ivoire in West Africa, he and his mentor Christophe Boesch have observed chimpanzees exhibiting a remarkably advanced set of tool-use behaviors. At certain times of the year when there are sudden windfalls of nuts, the chimpanzees will gather in boisterous groups to break the nuts open by pounding them with stones or short, heavy bits of branch. At full spate the sound echoes through the forest like a cheerful carpenter's shop. Young chimpanzees slowly learn, over a period of three to five years, to join this rewarding activity by watching their parents and other elders. Remarkably, the same tools

174

are often used by generations of chimpanzees. And the behavior is probably cultural, for it seems to be confined to this one forested region.

Now, employing techniques of molecular biology, Gagneux has been able to detect other patterns of behavior that are normally hidden from human observers.

Chimpanzee groups in this West African forest are relatively small, each consisting of twenty to a hundred members. Were such groups to remain genetically isolated from each other for many generations, they would begin to suffer from the effects of inbreeding. This can be avoided, or at least postponed, if genes can be exchanged between groups. Even before Gagneux began his investigations, it was known that such exchanges can happen, since young females will often leave a group and join another some distance away. He was soon able to determine yet another, far less obvious, mechanism of gene exchange.

He studied one group of about fifty chimpanzees intensively and identified them all at the DNA level. This was done very elegantly, simply by collecting hair left behind in the nests made from tree boughs that the chimps built each night.*

Only a tiny amount of DNA can be extracted from the root bulb of a single hair, but it is enough for the PCR technique to work its magic. (You will remember that PCR is so powerful, it can amplify the vanishingly small remnants of thirty-thousand-year-old DNA from a Neandertal. Getting enough DNA from the root bulb of a recently plucked chimpanzee hair is child's play by comparison!) For his family studies Gagneux used bits of DNA from the cell nucleus that have a different kind of inheritance from mitochondrial DNA. Like most of our other genes, these pieces of DNA are inherited from both parents. Because of this, he could determine the paternity of each offspring. This determination would have been impossible by simple observation, since receptive females normally mate with several males.

To his surprise, he found that new patterns of DNA continually appeared among the babies born in the group. The babies carried half the genes of their mothers, of course, but the new DNA patterns from their fathers were different from those carried by any of the males

---

* Such hair collection is harder than it sounds. Gagneux had to climb high into the trees, negotiating a gauntlet of sweat bees, wasps' nests, and biting ants, in order to get those tufts of hair.

that he had typed. There was only one possible explanation: Females, even those who seemed well integrated into the group's social hierarchy, must often be sneaking away into the woods for brief trysts with males of other groups in nearby territories. Although he was never able to observe this behavior directly, the story the genes told was unequivocal.

He presented this remarkable insight into chimpanzee mating behavior at a small conference at the end of 1995. Those of us in the audience nodded our heads wisely and exclaimed over what an interesting way this was to avoid inbreeding. How naïve and unworldly we all were! For when he later published his work in the journal *Nature*, it was not the long-term genetic consequences of this behavior that excited headline writers in newspapers around the world, but the behavior itself—the deliciously naughty "adultery" of the chimpanzee females.

The inconvenient fact that chimpanzees know nothing of marriage vows—let alone the guilt associated with breaking them—played no role in the brief media frenzy that followed. Nor did anyone seem to make much of the fact that if the females were committing adultery, the males were perfectly happy to take part in this wicked behavior as well. It takes two to tryst, after all.

Despite this story's soap-opera aspects, in the long term the prevention of inbreeding is most important to the survival of the chimpanzees. The genetic exchange allows chimpanzees living in the wild to be remarkably free of any harmful genetic effects that might otherwise accumulate as a result of their small group sizes. This may help to explain a puzzle that we explored earlier: Even though chimpanzees seem always to have been so rare that no fossils of them have been found, they have somehow managed to persist for millions of years.

The protection conferred by this outbreeding behavior may not last much longer. Chimpanzees' habitats are being destroyed rapidly, and surviving groups are being dispersed ever more thinly as we humans invade their territory—often killing them for meat or to protect cocoa crops against raiding. This invasion will make it more and more difficult for surviving chimpanzee females to bring new genes into their groups, or to find new groups into which to migrate. Human intervention is likely to do to the chimpanzees in a few brief centuries what millions of years of small population size could not do: bring about the demise of all of their wild populations.

FLIRTING WITH EXTINCTION

The relatively plentiful fossil record of our own species suggests that human population sizes tended to be large and that our ancestors did not have quite the same imperative to avoid the effects of inbreeding. While people do of course mate outside their immediate social groups, hunter-gatherer societies do not seem to show such strong outbreeding behavior as chimpanzees.

The pattern of our genes, however, tells us that for some brief moments we too may have flirted with extinction. Had those moments been prolonged, we might really have gone extinct.

Humans and chimpanzees are still so closely related that both species have essentially the same number of genes. But small DNA differences have accumulated between virtually all of our genes and the equivalent genes of chimpanzees. We do not yet know which of these genes help to produce such a strong urge to outbreed in chimpanzees, or what differences between those genes and ours might contribute to a less overwhelming urge in us. But while we wait for these genes to be discovered, we can in the meantime detect some other important genetic differences that can tell us a great deal about our different evolutionary histories.

For some years many groups of scientists around the world have been examining in detail the variation in a small piece of mitochondrial DNA that is shared by humans and apes. (It is the same piece, by the way, that was recovered from that Neandertal who lived long ago in the Neander valley.) The pattern of variation in this fragment provides, among other things, clues to the last time when our own species may have had a really serious brush with extinction.

We can glimpse the history of a human or ape population by determining how many of these genetic differences have accumulated in the population. If a great deal of such variation has accumulated, the population must have been quite large far back into the past—because mitochondrial variants that appear by mutation can accumulate relatively undisturbed in a large population. Fewer of these variants will be lost by chance, as would have happened if the population were small.

Small population size, by contrast, is the enemy of genetic variation. It is striking that chimpanzees have gone to such extraordinary lengths to increase their population size through outbreeding. Nor are chimpanzees alone in this respect. Other species of animals and plants have also acquired a great variety of outbreeding mechanisms. Indeed, the prevalence of outbreeding in the natural world has led many evolutionists to suspect that genetic variation must in itself be valuable.

If a population is large and yet has little variation, the most reasonable explanation is that something happened in the past to reduce the population's size. Such an event is what geneticists call a *size bottleneck*.

The term is quite descriptive. A population that has been large for some time may occasionally undergo a temporary reduction in numbers—perhaps as a result of ecological change, disease, or the migration of a small part of the population into a new territory. If such a reduction is then followed by a recovery period, during which the population once again achieves large numbers, then no visible signs of the bottleneck would remain. But the effects of the bottleneck would still be detectable at the gene level, for the population would have less genetic variation than its current large size would lead us to expect. We can infer that such bottleneck events must have happened to animals as diverse as cheetahs and elephant seals, because these species currently have unusually small amounts of genetic variation.

Just as an X-ray of the body can reveal an old lesion that has healed without an external trace, a look at the genetic diversity of a population can speak volumes about what has happened to that population in the past. Examining DNA variation can be a kind of genetic X-ray, showing signs of such traumatic events in a population.

Figure 9–1 shows the family tree that Pascal Gagneux and I constructed, using almost twelve hundred different mitochondrial sequences from humans, chimpanzees, bonobos, and gorillas. Like an X-ray, it presents in vivid form the remarkable differences between the amounts of diversity in humans and in our close relatives. Some of these populations, including ours, show evidence of fairly recent "lesions," and some do not.

The DNA samples were gathered from lowland and mountain gorillas by Karen Garner and Oliver Ryder of the San Diego Zoo; from East and Central African chimpanzees by Phillip Morin of the Sequana Corporation and Tony Goldberg of Harvard University; from western African chimpanzees by Pascal Gagneux and Rosalind Alp; and from the isolated Central African bonobos (pygmy chimpanzees)

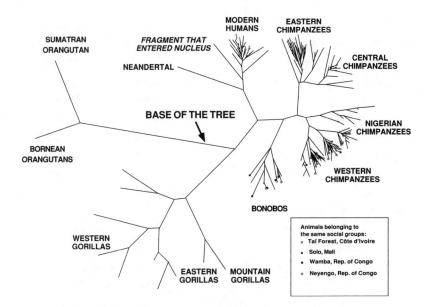

Figure 9–1. A family tree of 1,158 related pieces of DNA, taken from the mitochondrial chromosomes of our own species and our African and Asian ape relatives. The base of the tree is shown with an arrow. I have done a little editing of the data, to get rid of some noise.

by Ulrike Gerloff and her colleagues of the University of Munich. The human samples were gathered by many different scientists, most notably Linda Vigilant of Pennsylvania State University.

While they were collecting the data, the individual researchers had tended to concentrate on their particular species. But as Gagneux and I built the tree from all the sequences, some remarkable patterns emerged. We were able to see our own evolution very clearly in the wider context of the evolution of our close relatives.

We constructed the tree using an ingenious computer method invented by Naruya Saitou and Masatoshi Nei. In the tree, distantly related sequences are separated by long branches, while closely related ones form little twiglets at the end of a branch.

We can think of this tree as like a tree in a forest, with its trunk rooted firmly in the past and its branches pointed skyward toward the present. The tip of each twiglet represents an individual who is alive at the present time. The tree in the figure may be somewhat confusing, for its resemblance to a real tree seems to be minimal. But the branches were too numerous for us to draw them in a typical treelike shape viewed from the side. As in a real tree, some of the branches would have obscured, or been obscured by, other branches.

To avoid this problem, we have drawn the tree as a kind of star, with the branches radiating out from a common point (marked with an arrow). This shape makes sense if you imagine yourself suspended in space over the tree, looking down at it. From such a vantage point, all the branches radiating out from the base are clearly distinguishable.

The base of the tree clearly divides the orangutans from all the other groups. The branches of the rest of the species all tend to cluster on the other side of the tree, away from the orangutans. This is what one might expect, because by many measures orangutans are only distantly related to the other species on the tree. In effect, the orangutan branches have grown away from one side of the base, while those of the various other species have grown away in the other direction.

SQUEEZING THROUGH THE BOTTLENECK

At first the tree seems very complicated, but certain features emerge. Let us begin with our own species.

The human branch of this tree is straight and, except for the Neandertal sequence and another sequence that escaped into the nucleus from the mitochondrial chromosome a million years ago, is unadorned with twigs until the very end. This small and dense bundle of twigs, made up of more than eight hundred different sequences, contains representatives from many different human tribes and races, including a large number of African tribes, as well as Asians, Europeans, American Indians, Australian Aborigines, and New Guinea tribespeople. Although not all human groups are represented in the data, enough are present to give a good idea of the extent of human variation. It is unlikely that this tuft will change much, aside from getting even denser as more human groups are added in the future.

This tiny, dense human tuft stands in vivid contrast to the long, sprawling branches that lead to the gorillas. Soon after leaving the base, the main gorilla branch fans out immediately into a number of very long subbranches. Here and there, at the tips of the subbranches, are little tufts of twigs. The gorilla sequences within each of these tufts resemble each other very closely, but the sequences on the tips of the different long subbranches are only very distantly related. So far as their mitochondrial DNA is concerned, gorillas are enormously more diverse than humans.

Gorillas are found over a relatively restricted geographic region. Indeed, the very dangerous population decline that they are currently suffering is made even more serious by the fact that they were never common in the first place. Yet their genetic diversity may be even greater than we see here. The sequences shown in the figure are— because of the tremendous logistical problems involved in their collection—a rather poor sampling even of the few remaining wild gorilla populations.

Our guess is that gorillas have lived in isolated bands for very long periods of time, and with little genetic exchange among them. They really have flirted with extinction and have done so for millions of years. Like the chimpanzees, they have undergone little evolutionary change for a very long time. Since the gorillas at the ends of these long branches are similar to each other, most of the evolution that resulted in their gorilla-ness must have occurred during the time that passed between the base of the tree and the beginning of the divergence of the gorilla subbranches. These long subbranches tell us that gorillas have not done much evolving for at least the last four or five million years.

The bonobos are another isolated and highly threatened group of animals with a highly restricted geographic range, inhabiting only the forests of Central Africa south of the Zaire River. For decades these chimplike animals were occasionally captured and brought to zoos in Europe and America. There they were simply put in cages and allowed to fraternize with chimpanzees that had come from other parts of Africa. By the 1920s, however, careful anatomical and behavioral studies had shown that they really were different from chimps. Now they have been put into a separate species, *Pan paniscus*, which is clearly different from the true chimpanzees, *Pan troglodytes*. Their common name, *bonobo*, is an apparently African word of unknown origin.

Frans de Waal of the Yerkes Primate Center in Georgia, who has studied the behavior of bonobos intensively, recently summarized studies showing that they are significantly milder-mannered than *Pan troglodytes*. Some of them also seem more adept at learning words than the average *troglodytes*, though they may not be quite as good at tool use. Their sexual behavior is complex as well—they often use a wide variety of sexual encounters to smooth social interactions, and unlike *troglodytes* they often couple sexually face to face.

In spite of all these differences, however, bonobos are not some kind of chimpanzee–human hybrid (as has actually been suggested). They are really very similar to chimpanzees. This is obvious from their general appearance and behavior, and it has been confirmed from studies of the sequences of their DNA molecules and of the shape and number of their chromosomes. So few of them remain now, from such a restricted region, that their branch may be the only survivor of several bonobo branches that formerly existed. In the unlikely event that they are left alone in their forest and are allowed to migrate to other regions, they may regain that lost genetic diversity, though they will probably continue to be bonobos or something very like them.

The chimpanzees in the tree show yet another kind of pattern. They are currently far more widely distributed geographically than the gorillas or bonobos, and are found across a great swath of Western, Central, and East Africa. Although the sprawling chimpanzee branches in the tree reflect this dispersal, their branches are no more diverged than those of the gorillas. The mitochondrial sequences of the eastern, central, and western chimpanzees are clearly separated from each other; the western chimpanzees themselves are divided into complex subbranches. The ends of some of the subbranches have fairly dense tufts of twiglets. One particularly dense tuft, which looks very much like the one on the end of the human branch, is seen in the eastern chimpanzees. Most of the chimpanzees that contributed their DNA to this tuft live in Tanzania and include those that Jane Goodall has studied for three decades in the Gombe Reserve.

These various groups of chimpanzees are quite similar in appearance and behavior. They have all (after many taxonomic misadventures) now been put into the same species, *Pan troglodytes*. Although *P. troglodytes* has in turn been divided into three different subspecies, these groups are based primarily on geographic location and not on any consistent physically distinguishing characteristics. If we were to put an assortment of these chimpanzees into a police lineup, as we imagined ourselves doing in the Introduction with the orangutans, not even experts would be able to determine their origin with any consistency. To tell them apart, it is necessary to examine their genes.

The chimpanzee–bonobo split in the tree probably happened about three million years ago, and the various major branches of the chimpanzees parted from each other at least two million years ago. Thus the eastern, central, and western chimpanzees went their separate ways not very long after the chimpanzee–bonobo split. It astonished us, when we first saw this huge sprawl, to realize that these very different DNA sequences belong to members of a single species.

The little hollow squares in the figure, scattered across the western chimpanzee subbranches, represent members of the little group of chimpanzees that Pascal Gagneux studied in the Ivory Coast forest. Even this tiny group contains a huge diversity of different mitochondrial types, which have been separated for well over a million years. The mitochondrial chromosomes have recently been brought together in this single band in the forest, adding to its genetic diversity, by the laudable and untiring efforts of the females who migrated into the group. Even within Gagneux's little group, the genetic differences from chimpanzee to chimpanzee can be far greater than those that separate humans living in the farthest-flung parts of the old world.

The tree illustrates the enormous power of the outbreeding behavior practiced by female chimpanzees. Through their labors they have managed to bring together, in one tiny group in one part of a forest, mitochondrial lineages that long predate the appearance of *Homo sapiens*. Such dramatic genetic mixing has not happened with the gorillas, and it has certainly not happened with humans.

The most likely explanation, though not the only one, is that our own ancestors were squeezed through a size bottleneck, while the chimpanzees as an entire species were not. When this bottleneck took place, and whether there might have been a series of bottlenecks, remains to be discovered. But that something unusual happened to our ancestors seems unquestionable.

We are not unique in experiencing a bottleneck. Look again at that little tuft of sequences representing the eastern chimpanzees: it is almost indistinguishable from our own tuft. These eastern chimpanzees, too, may have gone through a size bottleneck.

Our best guess is that, some hundreds of thousands of years ago, a few chimpanzees managed to travel from Central Africa across the western part of the Great Rift Valley, possibly almost impassable in those days, into what is now Tanzania. There they quickly established a new population. As they passed through this size bottleneck, they lost genetic variation, but as the new population became established, they slowly gained it back. Nothing much else happened, however. The scenery in their new territory, with its vast valleys and mighty volcanoes, was certainly spectacular, but chimpanzees tend to pay little attention to scenery. For them, little had changed. Chimpanzees they were, and chimpanzees they remained.

Our little tuft of twigs, in contrast, stands alone. The subbranches that led to the various Australopithecines, to *Homo habilis* and *Homo erectus*, and most recently to the Neandertals, have all disappeared, and at this writing only one Neandertal sequence and an ancient sequence that managed to escape into the nucleus are left. If we did go through a size bottleneck, then our lineage managed to survive when all those others did not. And unlike the various chimpanzee subbranches that have lasted down quite nicely to the present time, the hominids at the tips of those lost subbranches were clearly different from us. Unlike the chimpanzee branches, our lineage has undergone enormous evolutionary changes and many of our close evolutionary companions did not make it.

SWEEPING CHANGE

Although our ancestors probably experienced a size bottleneck, we cannot now tell its cause. It may have been some chance event, perhaps a migration into a new territory, as with the East African chimpanzees; or it may have reflected some desperate time in our past, when a few of our ancestors were forced to adapt to new environmental circumstances. Such rapid adaptation would have drawn on the genetic variation present in the population in those days and could have led to huge changes in our gene pool.

A possible consequence of such an environmental change would be what population geneticists call a selective sweep. In such a sweep an advantageous allele spreads rapidly through a population and replaces the existing alleles.

Suppose that, hundreds of thousands of years ago, a group of our ancestors found themselves in a new environment that required rapid adaptation. Formerly rare alleles of many different genes would now be catapulted into prominence. The individuals who possessed these alleles would leave many progeny.

Over the next few generations these alleles would sweep through the population. If the sweep were rapid enough, even an evolutionarily irrelevant piece of DNA—for example, one of the population's mitochondrial chromosomes—might be carried along for the ride. It would replace the other mitochondrial chromosomes in the population, even though it had little, if anything, to do with adaptation to the new environment.

If the new environmental stress resulted in reduced population size, selective sweeps would become more probable. Formerly rare alleles can sweep more quickly if the population is small, and in order for a selective sweep to be really effective, speed is of the essence. If the sweep does not happen quickly, the genes being dragged along with the selected alleles will have time to lose their association with them. They will tend to be unmounted from their positions nearby on the chromosomes, through the processes of genetic recombination and assortment.

We do know one thing about any size bottleneck, with or without an accompanying selective sweep, that might have happened among our ancestors. It is difficult to see how it could have taken place if they had already dispersed themselves over a wide area. Suppose they had spread through Africa but had not yet fanned out across the Old World. Africa is a huge and complex place, with many different geographically dispersed habitats. Once hominids had spread through several of them, no single disaster or other event could have affected all of them simultaneously.

The bottleneck must have affected some hominid group that inhabited a single small region, almost certainly somewhere in Africa. Afterwards the survivors must have spread through Africa, displacing and perhaps driving to extinction all the other hominids that lived there. Then they broke free into Asia and Europe, perhaps in their bloodthirsty way to do the same thing there. The mitochondrial sequences show some signs of a second bottleneck that accompanied the migration from Africa into the rest of the Old World, but it is not as "clean"—there are genetic traces of subsequent migrations back and forth among African, European, and Asian populations.

ALL ABOUT EVE

The bottleneck question—whether one actually happened—is inextricably linked with a great scientific debate that is currently going on about human origins. This debate has to do with the age of a particular lady—something that is of course generally not discussed in polite society. Scientists, however, are not inhibited by such restraints and have spent years of intense argument over the age of the lady in question. She has become known as the mitochondrial Eve.

The Eve was first named a decade ago by Jim Wainscoat of Oxford University. He was careful to define who she was. Because mitochondrial chromosomes are passed down the female lineage without recombination, they must have had an ultimate source. That source was the mitochondrial Eve. She carried a particular mitochondrial chromosome from which all the mitochondrial chromosomes carried by humans living at the present time are descended.

But she was not—repeat, not—the equivalent of the Eve in the Bible, a woman from whom we are all descended! I emphasize this point because, while Wainscoat was careful not to make this mistake, both scientists and science writers in the decade since have repeatedly stated that the mitochondrial Eve, like the biblical Eve, was our sole ancestress. Wainscoat realized that the mitochondrial Eve had companions, and that both she and her companions bequeathed many other genes to us. Those genes were on nuclear chromosomes and not on the small mitochondrial chromosome. Although she was indeed the ultimate origin of all the mitochondrial chromosomes in our species, she played only a minor role in passing down the overwhelming majority of our genes.

The debate about where and when Eve lived is ongoing and inconclusive, but we can say some things here.

We know, from Krings's Neandertal DNA sequence, that Eve lived sometime after the split between humans and Neandertals. Further, if a bottleneck or sweep took place, then it must have done so at or after the time when the Eve lived.

We also know that Eve was probably not particularly remarkable. The fact that she carried the ancestor of all our mitochondrial chromosomes was after all a statistical accident, because of the way these chromosomes are inherited. She might have lived at some interesting time in our past, such as a period during which a bottleneck took place, but the chances are slim. After all, she was just one of our ancestors, most of whom lived in times that, although fraught with many dangers, were (luckily for them) not unusually so!

Mathematical theory and computer simulations show that the Eve could easily have lived long before a size bottleneck. She might, for example, have lived in a small tribe that managed to retain, for thousands of generations after she died, a few different mitochondrial chromosomes. Then, after hundreds of thousands of years, all but the descendants of her chromosome were lost, and the tribe fanned out in a sudden population expansion that permitted more chromosomal types to accumulate. There is no way to distinguish this possibility from the possibility that the Eve lived near the time of the bottleneck.

Thus the time when Eve herself lived cannot be settled at the moment. Most calculations have placed her at a date of between one and two hundred thousand years, at about the time of the appearance of early modern humans in southern Africa and the Middle East. But even a small change in the assumptions underlying these calculations will produce a much older date. By making some changes that I thought were justified, I recently concluded that she might have lived half a million or more years ago, and a few other scientists have arrived at similar figures.

All these are estimates of the median age of Eve, the most probable age. To add even more to the uncertainty, the errors on these estimates, particularly at the upper boundary, are so immense that we cannot be confident about the figure to within a few hundred thousand years! Eve, like any proper lady, remains coy about her age.

Nonetheless, combining evidence from molecules with evidence from bones makes things a bit clearer. The molecules tell us that we have far less mitochondrial DNA variation than chimpanzees do. So far, they are silent about when, how, and why we lost any earlier variation. The bones, on the other hand, tell us that many dramatic things really did happen in the course of our evolution, and we must always keep them in mind.

It would certainly be a mistake to attribute too much of our evolution to bottlenecks. Modern humans differ so much from our ancestors in so many ways that our species must have been shaped by far more than a single event such as a size bottleneck, with or without a selective sweep.

Popular science books tend to attribute the entire process of evolution to a single phenomenon—waves of cosmic radiation or giant asteroids from outer space, for example. But such an approach tells only part of the story. Although bottlenecks, and whatever caused them, may easily have been important, the genes that managed to squeeze through them were even more so. It is these genes that made us what we are and that set us on our new evolutionary path, so different from the evolutionary stasis of chimpanzees, gorillas, and orangutans.

AN EVOLUTIONARY TIME-WARP

We can tell many other things from these little bits of mitochondrial DNA. One of them gives at least a partial answer to the question: were the chimpanzees really chimpanzees all the way back? Have chimpanzees really changed hardly at all, even though during all those millions of years our own ancestors were undergoing one dramatic transformation after another? The fact that our own remote ancestors were very chimpanzeelike does not guarantee that the ancestors of chimpanzees were chimpanzeelike as well—they might have been quite different. Remember that we have no fossil evidence about those ancestors.

The bonobos provide us with the evolutionary yardstick that we need to answer this question. The common ancestor of the chimpanzees and bonobos must have closely resembled both species. If we knew when that common ancestor lived, then we could determine the minimum amount of time during which chimpanzees have been chimpanzees rather than something else. If the two species split recently, then we would have to guess what their common ancestors might have looked like during the millions of years before the split took place. But if they split much longer ago, closer to the human–chimpanzee divergence, then both the bonobo and common chimpanzee lineages must have been chimpanzeelike for all that time.

We can get a rough idea of the date of this common ancestor by comparing two numbers. The first is the amount of DNA divergence that has accumulated between the chimpanzees and bonobos, and the second is the amount that has accumulated between chimpanzees (or bonobos) and humans. After some statistical correction the ratio of these numbers shows that the chimp–bonobo split occurred about sixty percent of the way back to the human–chimpanzee split. This means that they probably diverged about three million years ago.

During those three million years many different evolutionary events must have taken place in these diverging chimpanzee and bonobo lineages. Yet none of them were so dramatic as to efface the essential chimpanzee-ness of either lineage. Indeed, as I mentioned earlier, before and even after the differences between chimpanzees and bonobos became known, they had been mixed together in zoo cages. Some of them readily mated with each other and produced hybrid offspring. The handful of these hybrids that we know about seem to be healthy, and there is no reason to suppose that they would be unable to have babies of their own. As with the hybrid orangs, however, it is unclear how well they might do if they were reintroduced into the wild, or whether subsequent generations will continue to be normal.

Although these different evolutionary lineages have now been brought together again in zoos, they pursued their separate courses for an even longer period than the Bornean and Sumatran orangutans. Indeed, at the time that the chimpanzee and bonobo lineages separated, our own African ancestors were all Australopithecines. Even *Homo habilis* had not yet appeared. It is remarkable that, even after all this time, gene flow can still take place between these species.

The ranges of chimpanzees and bonobos currently do not overlap; but what might happen if they were to meet in the wild? Would they mate, would they fight, or would they simply ignore one another? If it turns out that they do not mate in the wild, then they will have proceeded farther along the speciation process than the zoo hybridizations would indicate.

Let me emphasize once again that during that same three million years, our own ancestors changed dramatically. They acquired enormous brains, a truly upright posture, flexible and sensitive hands, the ability to construct elaborate cultures, the physical and mental equipment to be able to communicate through complex languages, and many other things.

In other words, since the time of the split between us and the chimpanzee–bonobo lineage, we have changed and they have not. The contrast is a vivid illustration of the uniqueness of our own evolutionary history.

THE TWO FACES OF SELECTION

As if these huge differences between ourselves and the chimpanzees were not enough, mitochondrial DNA provides still further clues to the uniqueness of our history. The clues, which seem at first to be paradoxical, arise from the nature of this region of the mitochondrial DNA itself.

The piece of DNA that Pascal Gagneux and I used to build the tree is not a gene in the usual sense. The part of the mitochondrial chromosome from which it is taken does not code for proteins. Instead, it is a necessary structural part of the chromosome, where the enzymes that duplicate the mitochondrial chromosome first attach themselves to the DNA. The reason that this particular piece of the chromosome has been used in evolutionary studies is a purely utilitarian one—it tends to accumulate mutations at a very high rate. Some of these mutations survive, and the resulting rapid evolution means that even closely related people or closely related apes tend to fall on separate little twiglets of the evolutionary tree.

The exact sequences of such pieces of DNA appear not to matter very much, since this type of DNA changes so quickly. But it does have a pleasing property that lends itself to mathematical investigation. Because the DNA is free to accumulate mutational changes, the rate at which it diverges from its relatives should be quite constant. Such a piece of DNA should act like a little molecular clock, ticking off the millennia in a regular fashion. So when we measure the amount of divergence that has accumulated between such sequences in different evolutionary lineages, we should be able to calculate the time at which the divergence between the lineages took place.

This supposed property is what allows us to measure the age of the mitochondrial Eve. But Pascal Gagneux, Tony Goldberg, and I found that while such clocklike behavior seems to be true for some kinds of DNA change, it is dramatically not true for one kind of change in one part of this mitochondrial region. There is statistical evidence that since the very beginning of the lineage that led to our species, right from the time when *Ardipithecus* lived in ancient Ethiopia, the kinds of changes that make substantial differences to the structure of this piece of DNA have been selected against, and selected against very strongly. Whatever its cause, this selection has not happened, or at least not as much, in our great ape relatives. Once again we are unique, but this time our uniqueness lies in our resistance to anything but small changes.

I have been emphasizing up to this point that our own evolution has been proceeding pell-mell, and that much more has happened to us than to our nearest relatives, the chimpanzees. But selection has two faces: it can accelerate some kinds of change and retard others. In the evolution of our bodies and minds, we have been revolutionaries. But in this bit of our mitochondrial DNA, it seems, we have been extreme conservatives.

Figure 9–2 shows this contrast clearly. This pair of trees, much simpler than the earlier tree, shows the number of changes in this piece of DNA that have happened in us and in our nearest living relatives. Because these trees are so simple, I have been able to make them look more like real trees, with our common ancestor at the base and with branches rising through time to the present.

The "transition tree" at the top shows changes of a type that have small effect on the DNA. These changes, technically known as transitions, occur when one DNA base has been substituted for another with similar chemical properties. Summing these changes gives the result that we, the bonobos, and the various branches of the chimpanzees are all about equally far from the gorillas. Since the time of that ancient split between the gorilla lineage and the rest of us, the transition clock has ticked along quite evenly.

This is emphatically not the case with the lower "transversion tree," which shows another type of change that has accumulated incrementally over the same time. These changes, called transversions, have a larger effect on the DNA. (The diagram to the right in the figure shows the difference between transitional and transversional changes.) In transversional changes, humans have lagged far behind. The lag has affected our entire branch, apparently right from the beginning.

This lag is continuing right down to the present. Some pairs of sequences taken from different people are separated by many transitional differences, but these same pairs show far fewer transversional differences than one would expect. At this very moment in our own species, transversional changes in this bit of DNA are being selected against.

The transversion slowdown seems confined to this one bit of DNA. The piece that lies right next to it does not show the effect—in this nearby region, transversions have accumulated as briskly down the human lineage as they have down all the others.

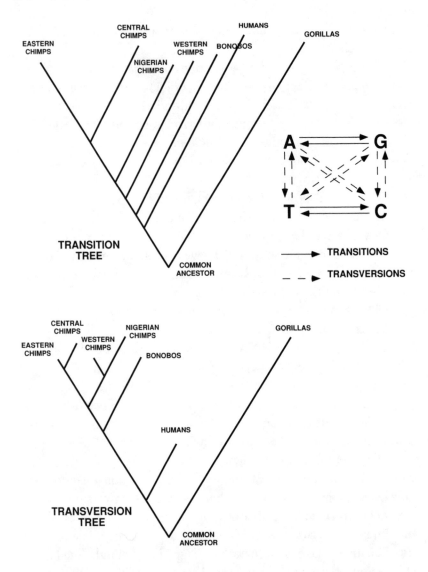

Figure 9–2. Trees showing how the transversion clock has slowed down for humans. The small diagram shows the difference between transitional and transversional changes in the DNA. The DNA bases A and G (adenine and guamine) are *purines*, larger than the *pyrimidines* T and C (thymine and cytosine). A change from a large base to a small one or vice versa appears to have a greater impact on the function of this DNA region than a change from large to large or from small to small.

What can be going on? Our current guess is that most transversion changes may damage the function of this important bit of DNA slightly and that this has been enough to select against these mutations. Many changes of the transversion type might interfere with the binding of enzymes to the DNA. This hypothesis is testable these days, and we are currently exploring ways to do so at the molecular level.

Mitochondria are thus very important to our survival, and not just any old mitochondria: human beings need tip-top mitochondria. The requirements for apes, it would seem, are a bit less stringent.

Throughout our ancestry, from *Ardipithecus* on, it seems to have been essential for our mitochondria to function at their very best. Conservative selection, which tends to weed out any changes that would do even the slightest harm, has been necessary to keep them at a peak level of functionality.

Several genetic diseases have now been traced to specific mutations in the mitochondrial chromosome. One of them is known by the mouth-filling name of Leber's hereditary optic neuropathy. As people with this condition grow older, a blind spot appears in their fovea, the highly acute central region of the visual field. The nerve cells in this region, which are among the most metabolically active in the entire body, begin to die because their mitochondria are not able to function at the extreme levels that are required. Mitochondria that have the same defect can continue to play their less demanding roles in other cells, including nerve cells, with no difficulty. If a fovea is to function properly, however, tip-top mitochondria appear to be needed. And if the requirement is to evolve big-brained, clever people like us, the demands on the mitochondria are likely to be stringent as well.

Why has it been more necessary for human beings than for chimpanzees or bonobos to preserve every ounce of the biochemical powers of their mitochondria? The big difference between us and the apes is the sheer size of our brains. Our brains have insatiable appetites—sixty percent of the energy consumed by a young child is funneled into brain metabolism and growth. Highly conservative selection, weeding out mitochondria with the slightest defect, appears to have been essential in order to ensure a plentiful energy supply to power revolutionary changes in our bodies and our brains.

Recently Jean-Jacques Hublin and his collaborators in Paris have obtained more information about the intensity of the selection that has shaped our brains. Their evidence comes from recent and fossil human skulls and from the skulls of chimpanzees. Using CAT scans, they measured the size of the sinuses through which the carotid arteries supply blood to the brain. The relationship between blood flow and brain size is a very direct one in humans and our ancestors; it is much less obvious in chimpanzees. If only the size of the sinuses is known, it is possible to predict very accurately the size of the brain in humans, but it is far less possible to do so in chimpanzees.

Human beings seem to be pushing the evolutionary envelope: if there is more blood flow, we can develop bigger brains. In chimpanzees, even if the blood flow is substantial, a bigger brain does not necessarily result.

The various phenomena that we have talked about here are all connected. The preservation of the base sequence of that small piece of mitochondrial DNA, and the strong relationship between blood flow and brain size, may be two aspects of the same thing. The overwhelming evolutionary imperative that has led to our large brains seems to have done much to shape both our genes and our bodies!

THE REST OF OUR GENES

This little bit of mitochondrial DNA has told us some amazing stories. It has illuminated our own history and shown us how different that history has been from that of our ape relatives. But this piece of DNA is very short, only four hundred or so bases long, which means that it makes up a mere one-hundred-thousandth of one percent of all the DNA in our cells. And as we have seen, it is rather unusual.

What have the rest of our genes been up to? They must have equally fascinating stories to tell. We are now beginning, dimly, to see what those stories might be.

TEN

❖

Sticking Out Like Cyrano's Nose

My nose is huge! . . . [L]et me inform you that I am
proud of such an appendage, since a big nose is the
proper sign of a friendly, good, courteous, witty, liberal
and brave man, such as I am.

A REMARK OF CYRANO DE BERGERAC, EMBROIDERED ON BY
EDMOND ROSTAND (1897)

THE WORLD OF DNA

Any DNA molecule can be represented as a string of bases of four
different types, arranged like letters in a sentence. Over time DNA
molecules tend to diverge from each other, as has happened with the
mitochondrial DNA of humans, chimpanzees, and gorillas. They will
diverge because they can become different from each other in so many
more ways than they can become similar again. This property allows
evolutionary trees to be constructed from DNA sequences.

Consider any sentence, even a short one, such as:

To be or not to be, that is the question.

Suppose this sentence is copied repeatedly by illiterate scribes,
who occasionally and at random substitute one letter for another. As
time goes on, the sentences will diverge more and more from each

195

other. I wrote a little computer program to act as such a clumsy scribe. After ten random mutations, the sentence looked like this:*

Td be vr nov to be, zhat im dhe quqstion.

After a hundred, it looked like this:

Fe st dp byy ar ta, tlxu xx nnj qmcpjztu.

Then I ran the program again, starting with a different random number. After ten mutations, I got:

Tt bl or not to bi, that it hhe question.

And after a hundred, I got:

Tw kl or gaa rp iz, awtp an asy uqadpwbi.

After only ten mutations in each line, it is still possible to trace out the familial resemblance. But after a hundred mutations, neither of the highly mutated sentences shows much resemblance either to each other or to the original sentence (though the word *or* was spared in the second computer run). Any further mutations would simply, in Churchill's phrase, make the rubble bounce. But I could have run the program any number of times and obtained a different nonsense sentence each time. DNA sequences are capable of diverging in many different directions.

An evolutionary tree built from DNA sequences has another pleasing feature. DNA molecules will continue to diverge even if the visible characters for which they code happen to be converging. Convergence of visible characters can often happen in the course of evolution. Consider the Ainu, an aboriginal group of people who are now reduced to living on the northern Japanese island of Hokkaido but who used to live throughout the Japanese archipelago. In appearance they differ greatly from the subsequent invaders of these islands—they have prominent facial features, an abundance of wavy brown hair, much body hair, and no epicanthic fold on their upper eyelid. Indeed, they are so Caucasian in appearance that anthropologists at first classified them in that group.

---

* I specified that if a mutation happened in a space between words, it left the space unchanged, to retain some recognizability in the sentence.

They are not Caucasian, however. Genetic studies show that the alleles of various genes that they carry are quite typical of the surrounding northern Asian populations. The Caucasian-like characteristics of their physical appearance seem to have arisen independently, though why they did so remains a mystery. (It should be pointed out, in order to avoid Eurocentrism, that the reason Caucasians themselves look the way they do also remains a mystery!)

This example illustrates vividly the dangers of making a tree using physical characteristics. If the characteristics have converged from independent origins, one can easily be fooled about their degree of relationship.

Because of DNA's property of divergence, evolutionary trees based on it tend to be more accurate, and more reflective of underlying processes, than traditional trees based on visible characters. Often it is easier to see the "true" evolutionary relationships among such species in a DNA tree than it is in a tree based on visible characters.

A typical gene tree for ourselves and our close relatives appears in Figure 10–1. It looks very much like the transition tree in Figure 9–2,

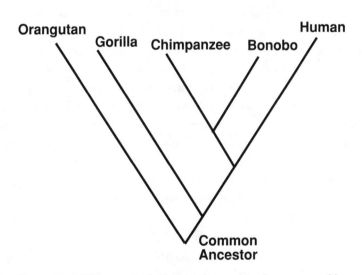

Figure 10–1. This tree was built from the combined sequences of four genes shared by humans and the great apes. The common ancestor occupies the bottom of the tree. It shows nothing unusual about humans—in this particular evolutionary sweepstakes, we have remained part of the pack.

in which the transitions seem to behave in clocklike fashion. There is no sign of anything unusual about the human branch; it is slightly longer than the others, but no more than one would expect by chance.

Surely, one would think, if human evolution really has taken off like a rocket, this fact should somehow be obvious in a DNA tree. Yet when a DNA tree is constructed, no matter which genes are used, the branch on which we are perched always blends nicely into the tree. Our branch is about as long as the branches of our nearest relatives. We really have to look very closely, as we did with the transversion tree in Chapter 9, to find any differences.

THE WORLD OF VISIBLE CHARACTERS

Once a month or so for the last two years, a group of scientists from the San Diego area have come together to discuss human evolution. This discussion group was the brainchild of my colleagues Ajit Varki and Rusty Gage. Members of the group hail from many parts of the huge biomedical research establishment that has grown up around the University of California, San Diego, campus, including the Salk Institute, the Scripps Research Institute, the Scripps Institution of Oceanography, and the San Diego Zoo. Our discussions have ranged widely, with intense debates about fossils, genes, primate behavior, and many other factors that have contributed to our evolution.

Already a number of collaborations are emerging from these discussions. One of them is a project that Ajit Varki and I have begun, to get a handle on a longstanding and extremely puzzling paradox about human evolution. Reduced to its essence, the paradox is this: The rates of evolution of our physical and mental attributes and of our genes simply do not seem to match. Our physical evolution seems to proceed much faster than our genetic evolution.

Varki and I realized that, unless we could explain this paradox, we could not explain human evolution.

In Chapter 9 I traced the story of how, from the time of *Ardipithecus* to the present, one very small but important piece of our DNA has slowed down in its evolution. But this slowing must surely be unusual. The many changes in our physical and mental capabilities during all this time strongly suggest that the rate of evolution of other pieces of our DNA must have sped up rather than slowed down.

Whatever genes control these characteristics must have changed at an accelerating pace in order to account for all the rapid evolution that is so plain in our fossil record.

Something else truly remarkable has happened to us as we diverged from the apes: Compared with those close relatives of ours, we can do so many different things! Alleles of various genes must have been selected for that contributed to that increase in the range of our capabilities. What might these genes be like, and indeed would we even recognize them if we came across them?

The heart of the paradox is this: When, as in Figure 10–1, we compare ourselves with our closest relatives at the DNA level, we simply cannot detect any sign of all this evolutionary ferment. We find no trace of this breakneck evolutionary pace in our genes—at least, not in the relatively small sample of genes that have so far been looked at in detail.

But in the visible world, the world of the phenotype, evidence for rapid change is abundant. Varki and I tried to build evolutionary trees that had nothing to do with DNA but rather were based on physical and mental resemblances. Trees of this kind hearken back to the early days of evolutionary studies, the dark ages when nothing was known about molecules. Darwin himself built such trees, and the relationships among living organisms that became apparent from them helped to convince him and other scientists of the reality of evolution.

The only difference between our trees and the ones that Darwin and his contemporaries constructed was that we built ours using up-to-date methods. These methods allowed us to see a very striking relationship indeed.

For years, Varki had been putting together a list of the many differences and similarities between modern humans and the great apes. The full list is enormous and highly elaborated. He found that the number of characteristics by which we differ markedly from our close relatives is astonishing.

He and I began our project by choosing random sets of the characters from his list and constructing a tree. Like a DNA tree, this tree was arranged so that closely related organisms tended to be grouped together.

To build this tree, we picked twenty-two physical and behavioral characters, including the ability to speak, the ability to recognize oneself in a mirror, the ability to walk bipedally rather than quadrupedally, the degree of difference in body size between the sexes, the difference in size between the two halves of the brain, the amount of thumb mobility, the period over which the young are nursed, the amount of skeletal muscle strength, the presence of breasts in a nonlactating female, and a number of others. We assigned a number to each characteristic on a scale of one to ten.

We found that while humans and apes sometimes had similar numbers for a given characteristic, usually they were very different. These differences make the astonishingly rapid pace of evolutionary change in our species glaringly obvious. As you can see in Figure 10–2, our own branch sticks out from among the branches of our close relatives like Cyrano's nose.

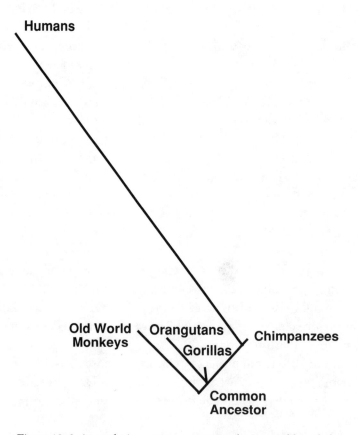

Figure 10–2. An evolutionary tree constructed using visible or behavioral characters. The base of the tree represents the common ancestor.

Note that the apes and monkeys differ among themselves in these characters. As one might expect, our closest relatives, the chimpanzees, are the nearest to us on the tree. But as we broke free of the pack, we have left even the chimpanzees far behind.

Sometimes, as in language ability, humans score high relative to apes and monkeys. Sometimes, as in skeletal muscle strength, we score low; the muscles of chimpanzees are three times as strong as ours. For most of these characters, however, we are clearly different from our close relatives, and when they are summed, the differences become enormous.

This tree is the mirror image of the DNA tree at the end of Chapter 9, in which humans lagged behind the apes in transversional DNA changes. A comparison of the two trees demonstrates very clearly the enormous disparity between the small amount of change that has happened to at least some bits of our DNA and the huge amount of change that has happened to us physically and mentally.

JUST HOW UNIQUE ARE WE?

Ajit Varki and I are hardly the first to have noticed that humans are so dramatically different from other animals, though so far as we know ours is the first attempt to put hard numbers on the differences. So great and obvious is this gap that for millennia people took for granted that our species resulted from some special creation event.

The special creation idea was sorely battered a century and a half ago, when Darwin used his theory of evolution to bridge the immense gulf between ourselves and the other animals. With one flash of insight, he showed that we really are part of the rest of the living world.

But old ideas die hard. Many people, including many scientists, felt strongly that Darwin's theory did not explain everything. He did not, it seemed to them, completely close the gap between ourselves and the rest of the animal world. Darwin's contemporary Alfred Russel Wallace, who had independently stumbled on the essential idea of natural selection, could not believe that human evolution was entirely the result of such an automatic process. In Wallace's eyes, the gap between ourselves and the other animals was still there, and it was as inexplicable as ever. He felt that some other, perhaps supernatural force must have been involved. Yet he could not explain the nature of that supernatural force, nor why we as a species were singled out for its benefits.*

In this book, as I hope you will have noticed, we nowhere invoke supernatural forces to explain human evolution. It can be explained by well-understood and perfectly ordinary evolutionary processes. What is unique is the speed and the power with which these processes have acted on us.

Long after Darwin, and after most scientists had abandoned special creation as an explanation, other aftereffects of the assumption of human uniqueness persisted. During the early part of this century paleontologists who attempted to reconstruct the history of our species were hampered because the fossil record they had access to was far more fragmentary than the one we know today. As a result they, too, were led astray by our obvious uniqueness.

The conventional view at the time, espoused most strongly by the great paleontologist Henry Fairfield Osborn, was that the human lineage and those of the great apes must have diverged very early. Osborn surmised that we probably originated in Asia, and that we and the apes must have embarked on separate evolutionary paths some thirty million years ago. This huge span of time was surely necessary in order to explain how we have become so different from our relatives.

The idea of an Asian origin was soon abandoned, but the idea of an early split was widely accepted by anthropologists right down to the 1960s, when it was overthrown with the advent of the science of molecular evolution. The chemists Linus Pauling and Emil Zuckerkandl of Stanford University were the first to suggest that molecules have evolved at a surprisingly constant rate and can be used as an evolutionary clock.

A molecular clock, it was soon realized, could be used to date the divergence between humans and chimpanzees. If the difference between the proteins of humans and chimpanzees was large compared to the divergence among mammals as a whole, then the split must have happened a long time ago. But if the difference was small, then humans and chimpanzees must have had a recent common ancestor.

* Wallace's idea was resurrected by Arthur C. Clarke and Stanley Kubrick in the film *2001*. The mysterious black obelisk that altered the course of human evolution in that film is nothing but Wallace's supernatural force, brought up to date and attributed to a galactic civilization rather than a deity.

Vincent Sarich and the late Allan Wilson of the University of California, Berkeley, were the first to carry out these measurements. Their data showed quite clearly that humans and chimpanzees have pursued separate evolutionary paths not for thirty million years but for a trifling five or six million. Their daring claim, at first resisted furiously by anthropologists, has now been proven correct by measurements on a wide variety of molecules in many different laboratories, and it has been confirmed, as we have seen, by new discoveries in the fossil record.

Once again, a worldview has been shattered, this time that of the anthropologists who had speculated about human origins. How much intellectual and scientific history is encapsulated in that little evolutionary tree of Figure 10–2! The immense physical and behavioral gap between ourselves and the apes seems to have led us badly astray at least twice.

Does the gap itself really exist? We are, after all, a terribly self-important species; perhaps we have simply exaggerated small differences in order to make ourselves seem unique. This possibility of bias, conscious or unconscious, poses a real problem for evolutionists. For just this reason many of them have never been satisfied with evolutionary trees of the kind that Ajit Varki and I constructed that are based on visible characters.

An enormous amount of subjectivity, after all, is involved in creating such trees. Why, for example, should we have picked one particular subset of visible characters out of the thousands by which humans differ from the apes? Did we load the dice too much by picking characters that are obviously different? Did we simply find what we expected to find?

We think not. Varki's list is made up of physical and behavioral characteristics by which we often, though not always, differ from the apes. In the majority of cases the variation from one ape species to another or from ape to monkey is dwarfed by the differences between these primates and humans. These differences are particularly striking for behavioral characteristics such as language and tool-making. Unfortunately these are just the characteristics that are most subjective and thus most prone to bias.

To get around this problem, we constructed another tree using only directly measurable physical characters and avoiding behavioral ones. Even after restricting ourselves in this fashion, however, our species still stuck out like that famous nose.

Bias is impossible to avoid completely, though I suspect that it is built into the way that any intelligent creature would look at itself and its near relatives. The composition of Varki's list was driven by the fact that humans happen to be available for comparison. I can illustrate this point with a thought experiment.

Suppose that our world is visited by a saucerful of utterly alien biologists from another planet. These aliens have a completely different body plan from our own—they resemble, say, intelligent insects. On their arrival they discover colonies of great apes and Old World monkeys, but for some reason they find no humans. (Highly unlikely, but since this is a thought experiment, I can postulate anything I please.)

The aliens study these primates by classifying them according to their visible or behavioral characteristics. As they do so, it will probably never occur to them to employ such features as bipedal gait, fine motor skills, or the ability to communicate through language. Such abilities will be only dimly or fleetingly apparent among these apes and monkeys. As a result, their list of characters will probably end up being much shorter, and far less varied, than Varki's list.

But if these aliens had been able to observe humans along with apes and monkeys, their list of characters would have been much longer. The reason is that, in the course of our own evolution, humans have taken to an extreme many of the slight and evanescent abilities exhibited by the other primates. At the end of such study, the aliens would almost certainly reach the same conclusions as ours. In any physical- and behavioral-character tree that they might build, the human branch would stick out dramatically.

BARKING UP THE WRONG EVOLUTIONARY TREE

The writer and cartoonist James Thurber loved to draw pictures of bloodhounds. His friend Hendrik van Loon, enraptured by the pictures, actually went out and purchased a bloodhound. But van Loon found the dog something of a disappointment. He later told Thurber, "That dog didn't care a damn about where I was. All he was interested in was how I got there."*

* "How to Name a Dog," *The Beast in Me and Other Animals* (New York: Harcourt, Brace, 1948).

Evolutionists who construct DNA trees are like van Loon's bloodhound. Such trees do indeed provide an overall picture of the shape of evolution and show us the proper evolutionary relationships among groups of species. But these trees cannot distinguish important changes from unimportant ones. The old-fashioned trees that are based on physical characters might be better for making such a distinction.

At the DNA level, as far as a tree-builder is concerned, all changes are given equal weight, even though most of them have little or no impact on the individuals carrying them. Unbeknownst to the tree-builder, only a tiny subset of these changes are the really important ones.

The consequences of tracking down this subset will be immense, for they are the very characters, at the DNA level, that contribute to the essence of humanness—or of chimpanzeeness. When we find them, we can measure accurately how these changes are continuing to accumulate among humans or chimpanzees at present.

When we reach that point we will not be confined to odd little bits of DNA like that mitochondrial sequence we spent so much time on in Chapter 9. We will be looking at the very stuff of our evolution itself. And we will finally be able to provide a firm answer to the central question posed by this book: Why are we evolving so quickly?

IMPORTANT GENETIC CHANGES

Sorting out the important from the unimportant changes in our DNA will soon be possible. The Human Genome Project, that vast scientific effort to sequence all our genes, is currently well on its way to realization. One of the many exciting prospects it opens up is that, over the next few decades, we will be able to compare all our DNA with that of chimpanzees and gorillas and thereby track down the many differences.

The substantial work that has been done so far has already revealed some surprising things about those differences. One is that humans are much closer to our near relatives than we might think, or than even sophisticated measurements tell us. It is often stated that we share 98 percent (or sometimes 99 percent) of our genes with chimpanzees. This statement is not right, and the error arises from confusing two rather different things, DNA sequences and genes.

The 98 percent figure really means that, when human DNA is matched with the corresponding stretch of DNA from a chimpanzee, we share 98 percent of the bases. But this is not the same as sharing 98 percent of our genes. Consider the gene at which sickle cell mutations arise, the beta hemoglobin gene. We have one of these genes, and so do chimpanzees. Even though the genes differ at a few bases, they are the same gene. Thus, by the criterion of whether both we and the chimpanzees have a beta hemoglobin gene, we are a hundred percent identical!

This is not surprising. Humans have been separated from chimpanzees for about five million years, but this period is less than 0.2 percent of the total span of 3.8 billion years that life has existed on the planet. By this measure, we should be more than 99.8 percent identical with chimpanzees—and indeed, so far as the number and types of our genes are concerned, we seem to be about this close.

The beta hemoglobin gene is just one of the hundred thousand genes we share with chimpanzees. In fact, despite some quite strenuous efforts, researchers have not been able to find any really "new" genes in humans—that is to say, any human genes that are completely different from those of chimpanzees. It appears that there has simply not been enough time, during our relatively brief separate evolutionary paths, for entirely new genes to evolve. We share not 98 percent but essentially a hundred percent of our genes with our close relatives.

To be sure, virtually all our genes are a little bit different from those of chimpanzees. In a few cases we have more copies of a gene than chimpanzees do, and there are a few cases in which chimpanzees have more. All our genes depart from those of chimpanzees by a few DNA bases here and there—a few letters in the sentences that spell out the genes' meanings. As a result, they are slightly different models of the same thing, rather like furniture that comes from different historical periods. A Louis XIV and a Louis XVI chair are both chairs, but they are distinctly different from each other. Yet most of these differences have little effect on their function—someone may perch with equal discomfort on both.

Like chairs from different historical periods, the same genes from different organisms also turn out to be quite equivalent. In some remarkable "gene swaps" made by molecular biologists over the past few years, genes taken from humans have been found to work very nicely in amphibians and even in fruit flies and bacteria.

But in order to work properly the genes must be brought under the control of the regulatory systems that are already present in those very different organisms. It is here that the differences between humans and chimpanzees chiefly reside.

In a royal court the regulation of chairs is at least as important as their function. The factotum who decided how many chairs should be placed in which rooms of Versailles had a great influence on the palace's livability. Similarly, the regulation of genes can be just as important as their function. Some regions of DNA that lie close to genes influence when and where in the body the genes are expressed. These regions have a great influence on the sort of organism that results.

A small change in such a regulatory region is quite enough to turn a gene on or off. More subtly, a change may delay or speed up the gene's timing during critical stages of development, or alter its expression in different tissues.

Many examples of such changes, some with huge effects, have now been found. In a fruit fly, for instance, the insertion of a small piece of DNA near one important set of genes is enough to give the fly an extra pair of wings. Humans all make large quantities of an enzyme called tyrosinase, which manufactures the dark pigment called melanin that colors our skin, our hair, and the irises of our eyes. When the tyrosinase enzyme itself is first manufactured in melanocyte cells, a regulatory region of the DNA decides whether it will be active or inactive. In some of us the enzyme is produced in its active form, so that we make large quantities of melanin. But somewhere in the ancestry of Europeans and Asians, mutations arose in this regulatory region that caused most of the enzyme to be made in an inactive form.

People who carry these mutations manufacture very little melanin. If the enzyme present in the cells of Europeans and Asians had been made in the active form, the result would be copious quantities of melanin, producing skins and hair as dark as those of Africans and Melanesians.

This small genetic abnormality, found in the "superior" white races, has had vast and terrible consequences for our species. Yet if such a change were to be examined solely at the DNA level, without reference to its effects on the body, it would be indistinguishable from the overwhelming number of other genetic changes that are scattered everywhere in our genes and that have far smaller effects or no effects at all. This is why evolutionary trees that are built using DNA sequences, like that in Figure 10–1, look so different from the physical character tree in Figure 10–2. The really important changes are certainly there, and most of them have important effects on gene regulation, but they are buried in the noise.

A small number of such highly significant regulatory bits of our genomes must have been changing at top speed over the last few million years. We suspect, without direct proof as yet, that fewer such changes have happened in the chimpanzee lineage. Thus, even though at first sight the evidence from DNA trees might seem to argue against it, we really must be evolutionary speed demons. The trick will be to track down the changes that demonstrate this.

The small subset of genes that are really responsible for our evolution will not continue to elude us for long. In Part III of the book we will look for clues to their nature, clues that we can glean from our remarkable behavioral diversity and our growing knowledge of brain function. We will see how changes in our environment can actually accelerate the evolution of our behavioral diversity. And we will explore scientists' first tentative attempts to track down these genetic changes. Finally, we will look at what all this growing understanding of our evolutionary heritage might mean for our long-term future.

PART III

Selection for Diversity

ELEVEN

❖

Going to Extremes

At this campus, twelve Nobel Prize winners have
taught or studied from nine different countries. A half
century from now, when your own grandchildren are in
college, there will be no majority race in America. . . .
As you have shown us today, our diversity will enrich
our lives in nonmaterial ways, deepening our under-
standing of human nature and human differences,
making our communities more exciting, more enjoy-
able, more meaningful.

PRESIDENT BILL CLINTON,
UC San Diego commencement address (June 14, 1997)

One Saturday in 1997 my wife and I visited the X-games in San Diego. X stands for "extreme," and the games certainly lived up to their
name. This twice-yearly event, started in 1994 by the cable television
channel ESPN, celebrates newly invented competitive events. They
involve brand-new skills, such as in-line skating and bungee-jumping,
that did not exist a few brief years ago.

The hastily erected X-games park was filled with thirty thousand bronzed and healthy young people, watching a range of contests
that their parents would have found unimaginable. The games fall into
eleven categories, made up of twenty-seven different disciplines. While
most of the 450 athletes in attendance were male, about twenty percent were female and this fraction is growing rapidly. So are the

rewards—top athletes in the most challenging X-game sports are now earning more money through product endorsements than their Olympic counterparts. The athletes themselves are an unusual mix and include a number of professionals such as doctors and lawyers.

ESPN has managed to change the habits of the American television-watching public. When X-games are on television, they sometimes outdraw more conventional sports like football and baseball.

The most extreme of the extreme was the snowboarding competition. Although snowboarding is now an Olympic sport, these summer games took place before the Olympian transformation. Here the snowboarders used a hundred-foot tower, built approximately in the shape of a squat, wide ski-jump. The tower was festooned with cables along which the automated ESPN cameras could scoot, allowing them to follow almost the same trajectories as the athletes and catch every bit of the action.

That morning, in cheeky defiance of nature, workers had coated the surface of the plastic-covered jump with a layer of snow. By the time the event started, the snow was already melting briskly in the southern California sun. Between jumps, game attendants with shovels crawled here and there over the slope, trying to break up the worst of the icy patches.

Balanced on snowboards, which are essentially oversize skateboards without the wheels, jumpers hurtled down the upper slope. Sometimes they would use the board as a snowplow to brake their descent; sometimes they turned it parallel to the slope in order to hurtle even more rapidly toward the jumping-off point. Once in the air, they showed how far they had diverged from the staid ski-jumpers who formed part of the distant ancestry of this fearsome sport. They began to somersault, or to whirl around like tops. Then they divided into two groups. The first group met the ground in a less-than-graceful collision, cartwheeling to an eventual halt in a confusion of arms and legs. The second group, as casual as could be, finished their last spin or somersault exactly on time, landed with their snowboards in exactly the right position, and coasted to an elegant stop over the bumpy remnants of the snow.

The reader will not be surprised to learn that, confronted with a spectacle such as this, I immediately wondered about its evolutionary implications for our species. They are, I hope you will agree, profound.

SELECTION FOR DIVERSE CAPABILITIES

Ecologists divide species of animals and plants into specialists and generalists. Specialists, like the gorillas we examined in Chapter 8, are confined to narrow ecological niches. It is difficult to imagine that, if they were accidentally to be transported to another continent or even to another part of Africa, they would multiply uncontrollably and produce a plague of gorillas. But many other animals and plants—generalists—have done just that: starlings and house sparrows imported from Europe to America, rabbits brought from Europe to Australia, kudzu that has spread from Japan to remote parts of the world such as the Galapagos archipelago. Humans are perhaps the most successful generalists of all, and the world is currently suffering from a plague of humans.

Part of our success is due to our ability to modify our behaviors in the face of new environmental challenges, and to build on a knowledge base in order to do so. Snowboarding is a madly logical blending of skateboarding and ski-jumping, but the idea for it would have come only to somebody who was familiar with both sports. Such a sport would never have occurred to a Cro-Magnon, even though, judging by brain size, he or she had essentially the same mental capacities as a modern human.

In range and variety, the basic skills needed to survive in the Cro-Magnon world were not much different from those needed to survive in the modern world. It is a challenge to make useful tools from stone and bone. Hunting skills are complex and subtle, as are the abilities and knowledge that one needs to gather wild plants for food and medicine. The skills we currently need, while different, may actually be less difficult than those required of a Cro-Magnon. Being able to read at a minimal level, to do one's income taxes, and to drive a car are probably less demanding than being able to track and kill a deer in a winter forest.

But our world differs dramatically from the Cro-Magnon world in one important aspect: its sheer bewildering variety. Moving beyond the realm of what we *must* do to the realm of what we *can* do, we are confronted with choices and opportunities that grow with every passing year. And yet, no matter how outré the challenges may be, some of us will rise to meet them.

If a group of Cro-Magnon babies were suddenly transported to the present time, they could learn how to take advantage of many of the opportunities offered by our complex world. They would probably not exhibit the full range of possible behaviors—the chances are slight that such a small group could produce concert musicians, top-flight engineers, brilliant mathematicians, and expert snowboarders. But they might easily produce a star in one of these categories, and many of them would certainly prove moderately skilled in others.

All evolutionists would agree with this scenario, but they have always been puzzled by how Cro-Magnons—or indeed any primitive people—might have acquired such capabilities if their ancestors had never been exposed to such challenges. How could the genes for these capabilities have been selected for before the selective pressures existed?

The solution to this problem will give us important clues to how our brains have evolved and are continuing to evolve. But as yet we have a limited understanding about how our brains function. Some distressingly simple-minded explanations have been proposed: geneticists, for example, have attempted to build models of how our genetics might have been influenced by our culture and vice versa. They begin by postulating a straightforward one-to-one connection between genes and behaviors.

One of the most detailed of such models was made by Charles Lumsden and E. O. Wilson in their 1981 book *Genes, Minds, and Culture*. They suggested that particular features of culture, features that they called *culturgens*, might increase the survival of those members of the population who carry particular allelic forms of genes.*

Simple-minded though this approach is, on its face it seems not entirely unreasonable. The model is, after all, based on well-established evolutionary theory and on much observational data from animals and plants. Often there is a straightforward connection between a gene and its environment. We saw some vivid examples of this in Chapter 4 when we looked at the selective response of humans to malaria, which resulted in rapid selection for alleles such as sickle cell that confer at least some resistance to the disease.

Another clear example is one that we all remember from our high school days, the story of the moth *Biston betularia* and how it changed

---

* They were remarkably coy about defining culturgens, giving only an extremely general definition and virtually no examples beyond such vague things as "avoidance of incest."

as a result of the English industrial revolution. Effluent from factories killed the lichens that covered the tree trunks on which the moths habitually rested. The bare trunks were darkened still further with a coating of soot. Within a few years light-colored moths were replaced by dark-colored ones. But the environmental change did not cause the genetic change. Even in the days before industrial pollution, moths that were dark-colored had always been arising by mutation. The difference was that these dark moths had suddenly become advantageous.

If evolution worked entirely in ways similar to this moth example, then we might well suppose that particular forms of genes could be selected for by particular features of our culture as they were invented. But the moth example actually illustrates only one very simple kind of evolution. We have already explored some much more complicated evolutionary stories, in which the connection between the selective pressure and the result is nowhere near as obvious. Evolution works on the brain in ways that are very complicated indeed.

Geneticists wedded to the idea that each gene has its function tend to have a hard time coming to grips with this, but ordinary genetic rules do not apply in the world of the mind. For one thing, few if any genes are directly connected to culturgens. Specific genes for specific behaviors—like playing the piano, or playing the stock market—do not exist.

While this observation may seem trivial, it has not prevented geneticists from continuing to search for such genes. At various times over the last few years, groups of scientists have announced the discovery of genes that are claimed to explain a great variety of behaviors. Lately, the triumph of genetics spearheaded by the Human Genome Project has increased the number of such claims. But most of these claims, on closer examination, have melted away like the snow at the X-games. When we examine the actual details of the searches, we can begin to see why this has happened. And, as we proceed, we will begin to uncover clues to why this is so, and why the evolution of the brain has been so unusual.

HITTING THE BRAIN WITH A HAMMER

An obvious place to look for genes that influence behaviors is in people who are mentally retarded, since they show a loss of abilities and greatly altered behaviors. One might think that if we could pin down which genes cause retardation, we would know the genes for these abilities and behaviors.

Many people incarcerated in homes for the mentally retarded are there because of genetic damage. In most cases the damage is localized to a single gene, or a small region of a chromosome carrying a few adjacent genes, or a single chromosome. It is easy for genetic recombination to get rid of the damaged gene or chromosome—about half of all children born to people who are severely mentally retarded are of perfectly normal intelligence. Because many of these genetically normal children have extremely difficult childhoods, this fraction will certainly turn out to be higher as social workers carry out more early interventions. If retardation were due to some generalized damage that spreads throughout the chromosomes, then all or most of these children would be abnormal.

In general, the causes of retardation are not problems with the genes that control specific brain capabilities. Instead, the damage is caused by genes that affect the overall development of the brain or its overall function.

The common genetic causes of really severe brain malfunction can be counted on the fingers of one hand, although any one cause may claim hundreds of thousands of victims. For example, among all cases of severe mental retardation, regardless of whether they are caused by genes or by some environmental or developmental accident, fully a third are in fact genetic and can be traced to one particular genetic abnormality.

This abnormality is Down's syndrome. It results when, by accident, a fertilized egg ends up with an extra chromosome 21, so that each cell carries three instead of two copies of all the genes on that chromosome. Even though all these genes are likely to be perfectly functional, too many copies of them can be profoundly damaging.

Down's syndrome affects development in many ways. At least some of the associated mental retardation has been traced not to an abnormality of the brain itself but instead to a tendency for the pulmonary arteries, which supply blood to the lungs, to become blocked by excess tissue growth. Unimpeded blood flow is crucial to a developing fetus—so much so that a high rate of blood flow has been selected for in Tibetan mothers. The brain of a Down's baby is affected by this excess tissue growth taking place elsewhere in its body.

Although the brain of a Down's syndrome child is potentially normal, it cannot reach its potential because of a genetic disturbance that has nothing whatsoever to do with the brain's function per se. Later in life, other accumulating abnormalities, some with a striking resemblance to Alzheimer's disease, cause further damage, but the early arterial blockage triggered by the genetic abnormality helps to set the pattern of mental retardation.

A substantial number of non-Down's cases of mental retardation are due to a genetic condition called fragile X syndrome, in which a mutation causes an abnormal lengthening of a part of the DNA itself. This first mutation increases the likelihood of subsequent mutations, so that the lengthening can actually grow worse in succeeding generations. The syndrome becomes progressively more severe as it passes down through a family.

Lengthening this DNA turns off a nearby gene that seems to be an important part of the pathway by which information passes from the cell's nucleus to the rest of the cell. This obstruction appears to damage the metabolically active cells of the brain more than it damages the cells of the rest of the body. Unlike Down's, the abnormality does lie in the brain itself, but it is a generalized abnormality that causes generalized brain damage. In effect, the brain has been hit with a hammer.

Molecular biologists are now probing other less common genetic defects that lead to mental retardation. One of the most remarkable of these is Williams syndrome, a rare condition that happens about once in every twenty thousand births. It produces a striking set of symptoms.

People with Williams syndrome are frail and suffer from heart problems. Their faces are thin and their chins are pointed, giving them an appearance described as "pixielike." The irises of their eyes show a lacy texture. And they suffer from varying degrees of mental retardation, from slight to profound. Yet their retardation is odd and strangely selective.

Victims of Williams syndrome are cheerful, outgoing, and gregarious and often have remarkable musical talent. They love to tell complicated and highly imaginative stories and use an extensive spoken vocabulary. A Williams child might, when asked to name animals, come up with an amazing collection of real, mythical, and extinct beasts ranging from apatosaurs through unicorns to zebras.

Many Williams children imagine that they might become writers someday, even though most of them can only read or write to a limited degree. All of them show another strange deficit—they are unable to put together a pattern out of its component pieces. A drawing of a bicycle made by a child with Williams will consist of highly distorted bits of bicycle joined together in a seemingly random fashion.

This pattern, in which some mental abilities are apparently spared while others are damaged, has led to tremendous interest in the syndrome. Is Williams, which is clearly genetic, caused by damage to genes that code for some brain functions but have little influence on others? Are there, in short, genes for these specific behaviors?

A large group of researchers at the University of Utah and five other institutions have now untangled some of the genetics of Williams syndrome. The story is one of genes that interact with each other in dramatic ways, and sometimes do seem to affect specific behaviors.

The syndrome results from a tiny deletion of part of chromosome 7. The size of the deletion varies among different people with Williams syndrome. It is sometimes substantial enough to remove a dozen genes and occasionally small enough to remove only two. People heterozygous for this deletion have just one copy of each of the genes in that region, a copy that was carried on the undeleted chromosome that they inherited from their other parent. Note that this is the reverse of the situation with Down's syndrome—in Down's, the problems arise from having too many genes, while in Williams the problems come from having too few.

One gene that always seems to be included in the Williams deletion codes for a protein called elastin. This protein contributes to the elasticity and strength of arterial walls; its deficiency may explain the circulatory problems and the strange lacy appearance of the irises so typical of the syndrome. But by itself it does not explain the unique pattern of mental retardation, for some people have only one copy of the elastin gene but are missing none of the nearby genes. Even though they have the same circulatory problems, most of them are mentally quite normal.

It is the lack of the other genes that seems to contribute to the mental deficiencies. The Utah researchers examined carriers of very small deletions, who are missing one elastin gene and one other nearby gene. These people are not mentally retarded, though they have the physical appearance and circulatory problems typical of the syndrome. But they do express very strongly one behavioral characteristic of Williams: the inability to draw coherent pictures.

This second gene, rather than the elastin gene, is most likely the one that influences brain function. Unlike the elastin gene, it is turned on in various parts of the brain. It codes for a type of protein that is able to modify other proteins in the cell chemically, but nothing else is known about it. Why it affects brain function remains a mystery. Nonetheless, having too little of the protein must lead, through who knows how many developmental steps, to this remarkably selective mental deficit.

The gene is unlikely to act alone. To begin with, it may interact with the elastin gene. People who have two copies of the brain protein gene but are missing one copy of the elastin gene seem, for the most part, to be quite normal. But what about people who have the reverse situation—two copies of the elastin gene but only one copy of the brain protein gene? Team member Michael Frangiskakis tells me that they have looked for such people but without success. They may remain undiscovered among the many people who have slight learning deficits, or they may be completely asymptomatic, concealed somewhere in the population at large.

This story resembles the story of Leber's optic neuropathy that we encountered in Chapter 9. In that disease the harmful effects of the mitochondrial mutation are expressed only under the severest conditions, while the same slightly damaged mitochondria are able to function normally in most of the cells of the body. Similarly, people with Williams syndrome make enough of the brain protein in most of their nerve cells, but during critical times in development, some cells may be particularly liable to damage because they do not make quite enough of the protein.

Is this second gene a gene for a specific behavior? Perhaps, but far more likely, it is interacting with many other genes. The brain is too complicated an organ, the result of too many complicated gene–gene and gene–environment interactions, for specific functions to be controlled by single genes.

Many other rare types of mental retardation, with a wide variety of symptoms and names like Angelman, Langer-Giedon, and Prader-Willi, have been traced to little deletions on various chromosomes.

Some of these deletions remove a single gene, but most, like Williams, remove more than one. They cause a great range of symptoms, and one day, as with Williams, these symptoms will be traced to deficits of specific proteins. But these proteins are not likely to be the ones that control these behavioral deficits—instead, I would predict that when the proteins are made in reduced amounts, they will be found to damage complex functions of the brain that are each the sum total of many different genes acting together.

These various syndromes show that in order for one or two genetic changes to affect mental function, the changes must be severe enough to hit the brain with the biological equivalent of a hammer. Gene hunters can track down genes with such large effects easily, but tracing the many different genes that together produced these complex brain functions in the first place is far harder. The gene for piano-playing will continue to elude us.

TWELVE

※

How Brain Function Evolves

No one, I presume, doubts that the large proportion
which the size of man's brain bears to his body,
compared to the same proportion in the gorilla and
orang, is closely connected with his higher mental
powers. . . . On the other hand, no one supposes that the
intellect of any two animals or of any two men can be
accurately gauged by the cubic contents of their skulls.

CHARLES DARWIN, *The Descent of Man* (1871)

As we saw in the last chapter, there seems to be no obvious connection between specific genes and specific behavior patterns. But without such a connection, how could our brains, with their wide variety of abilities, possibly have evolved? Various alleles of genes that regulate brain development must have some connection with the environment that selects for them, for otherwise natural selection would have no effect. The connection is indeed there, but it turns out to be a subtle one. Genes and environment interact in such a complicated way that the whole process resembles the complex weave of a tapestry. Although it is difficult to follow the individual threads, in this chapter we will try.

THREADS OF THE GENE–ENVIRONMENT TAPESTRY

People with the condition called narcolepsy exhibit an odd and life-threatening behavior: They fall asleep suddenly and inappropriately.

An old joke has it that it is normal for an audience at a lecture to fall asleep, but you know you are in the presence of narcolepsy when the *speaker* falls asleep!

Jokes aside, narcolepsy is a very serious condition. Among Caucasians there happens to be a direct connection between this behavior and a particular gene, one of the strongest that has yet been discovered. Virtually all Caucasians who exhibit narcolepsy carry a particular allele in one part of a complex collection of genes called HLA, which is found on chromosome 6.

Merely having the allele does not sentence the carrier to narcolepsy, however. While almost all Caucasian narcoleptics have the allele, only a small fraction of those with the allele are narcoleptic. The allele is found in twenty-one percent of the Caucasian population, but we do not see one in every five of this group suddenly nodding off; only a third of one percent of those who carry the allele actually show the condition. A few narcolepts do not carry the allele, yet develop the disease nonetheless.

Having the allele, then, does not appear to determine one's fate with certainty but constitutes only one step on the road to narcolepsy. The story, when it is finally unraveled (which will likely take decades), will certainly turn out to be very complicated. But it is possible to speculate a little. The gene is located in the HLA complex, which is known to influence disease resistance and susceptibility; this suggests that some childhood episode of infectious disease might act as a trigger. Such an environmental event would presumably be fairly rare—perhaps an unusual disease or a particularly severe bout of some common disease. Those of us who do not carry the allele might easily have had the disease during childhood, but we would have recovered without long-term effects.

Is the HLA allele, then, the gene for narcolepsy? No, although in some as-yet-to-be-determined way it does contribute to the condition. The damage it does to the brain is not the result of the activity of the gene itself, but probably happens because carriers of the gene are more susceptible to some environmental stress; without that stress, nothing would happen.

Genes that supposedly contribute to behaviors in some fashion have become the new darlings of the media. Television and newspapers have recently trumpeted the discovery of genes "for" alcoholism, hyperactivity, and criminal behavior, among others—even a putative gene "for" divorce!

Undoubtedly some genes affect behavior. But in general they are not genes "for" specific behaviors. Consider the newly discovered gene "for" attention deficit hyperactivity disorder (ADHD), which recently received worldwide publicity. Children with ADHD have great difficulty concentrating on schoolwork and cannot sit still for long periods. Usually, but not always, they outgrow these symptoms. Ritalin is the medication of choice for ADHD, and more than a million children throughout the United States are currently on this drug.

Such wholesale prescribing may in part be a result of the way ADHD is diagnosed. The standard diagnostic test for ADHD lists fourteen different criteria, and a child with a high score in eight or more is considered to have hit the jackpot, so to speak. But a majority of male children in the population score high on at least a few of the criteria (females tend not to show such high scores), which means that ADHD is really something of an artificial construct.

The syndrome lies at one end of a continuum of behaviors. Drawing the line between ADHD and behaviors that are not quite ADHD is thus very difficult, and highly subjective. Because the drug is prescribed in such wholesale fashion, I suspect that all children in literature would have ended up on Ritalin if they were real and alive today, with the possible exception of Little Lord Fauntleroy.

The connections between specific genes and ADHD are tenuous but intriguing. It has been known for a long time that some hyperactive people simply do not respond to thyroid hormone treatment; in a few cases this refractoriness runs in families. Starting in the mid-1980s, work in many laboratories led to the discovery that the defect in these families lies in a gene that codes for a receptor protein that binds to thyroid hormone. The altered protein is unable to bind the hormone properly.

In 1993 a group at the National Institutes of Health examined eighteen of these families and found that over half the children with a defective gene fit the criteria for ADHD. Further, half the adults carrying the defective gene probably had ADHD when they were children. By comparison, thirteen percent of family members without the mutant gene either exhibited ADHD symptoms or had the condition when they were children, which is slightly but not significantly higher than the frequency of the condition in the general population.

The Associated Press headlined its story about this work "Defective Gene Produces Hyperactivity." This is, to say the least, overstating the case. The gene "for" ADHD actually has much in common with the gene "for" narcolepsy. A number of family members who carry the defective gene show few if any symptoms, and a number of those who do not carry the defect show many symptoms. Probably the defective gene is able to *affect* the level of hyperactivity, but it hardly *produces* the behavior.

The discovery that a defect in this one gene has a statistically significant impact on attention span and activity levels is nonetheless exciting, and much research is currently being directed toward the whole complex issue of thyroid hormone activity and its influence on these behaviors. But the discovery of this defect does not constitute the discovery of a specific "gene for hyperactivity." Rather, it tells us that a specific defect in the mechanism of binding of thyroid hormone to cells may exaggerate certain behaviors.

The thyroid hormone gene is not alone in its influence on ADHD: three different genes that are involved in binding or transmitting the important neurotransmitter dopamine have been discovered to be polymorphic in the human population. This is important because sensitivity to dopamine, and a similar compound serotonin, have been found to be correlated with a whole web of behaviors. Some of the various alleles of these genes have slight but significant associations with ADHD, including Tourette syndrome, stuttering, obsessive-compulsive disorder, substance abuse, and other psychiatric ailments.

It must be emphasized that in none of these cases is the correlation anywhere near perfect. These genes are a long way from directly affecting the mechanisms, whatever they are, that define this wide assortment of behaviors.

Yet these findings are unquestionably of enormous importance. As a result of these and similar discoveries, psychiatry is rapidly moving into a new and exciting phase that holds promise for specific palliatives and even cures for some extreme and damaging behaviors. But as exciting as these discoveries are, we are only just beginning to penetrate the first of many levels of brain function. It is as if, in the course of investigating how an automobile works, we discovered that increasing the richness of the fuel vapor entering the engine causes it to speed up. While this finding may suggest many other experiments, it tells us nothing about the engine itself.

Let us take a moment to define our terms. In studies such as these, the word *behavior* has a very specific meaning. When authors speak of "behaviors," they do not mean some aspect of intellectual function, but rather some thoughtless, driven, or inappropriate activity. A person who exhibits one or more such "behaviors" may at the same time be able to function intellectually at a high level.

Consider drug addiction. People who become drug addicts appear often to be driven toward that addiction by biochemical imbalances that produce emotional dysfunction, rather than intellectual deficits. We define *drug addiction* as a behavior, but it has nothing to do with the higher functions of the brain. Better to call it, as many psychiatrists and psychologists do, a behavior pattern, with the implication that it is caused by some global influence which distorts the function of the brain. A gene that influences such a behavior pattern is a long way from a gene for piano-playing.

Many such behavior patterns are unacceptable in our highly structured society; hence all of the crude chemical attempts to correct them. In some cases, our remarkable new armamentarium of drugs actually helps us succeed in modifying the behavior patterns so that they can conform more closely to societal norms. Children successfully treated with Ritalin and the newer generation of ADHD drugs are able to acquire skills that permit them to survive more effectively in our society.

As that society changes, acceptable behavior patterns change as well. These continually shifting societal parameters are a moving target for the selective pressures that act on us. We are also adding new and unexpected selection pressures: a child who cannot respond to Ritalin may grow up, carry out some act that is unacceptable in our current society, and end up incarcerated—and effectively removed from the gene pool—as a result. What is being selected against here: genes that predispose a child to an unacceptable behavior pattern, or genes that make a child unable to respond to the drug of the moment? It is impossible to tell—this thread of the tapestry cannot be disentangled from the others.

BENDING THE TWIG

The genetic connections to these behavior patterns are still tenuous and volatile. Yesterday's highly touted gene "for" this or that behavior pattern often turns out, after further work, to be an artifact, usually because subsequent studies are unable to replicate the original observation. This has not been the case for the associations of ADHD and other behavior patterns with alleles of dopamine receptor and transporter genes, as weak as these associations usually are; many of these findings have now been independently replicated in other laboratories. But a number of claims that genes have been discovered "for" such conditions as bipolar depression and schizophrenia are now known to have been premature.

Retractions of highly publicized discoveries rarely appear in newspapers or on television, so the public gains the impression that scientists are adding to an ever-growing list of genes "for" this or that behavior. In fact, they are doing nothing of the sort. What they are doing is finding a tentative list of genes that affect brain function, on some global or slightly more specific biochemical level. At the same time, they are discovering that these genes are influenced tremendously by environmental factors. Environmental effects are often so strong that they can cause the supposed effects of a gene to appear in one study and disappear in the next. If we are to begin to understand how genes for enhanced brain function can be selected for, we must understand the role of the environment, since it can both mask and enhance a genetic connection.

The impact of the environment is often overwhelming. Poor nutrition and disease can affect our mental capabilities, sometimes more than we suspect. One of many vivid examples comes from the annals of mental illness.

Schizophrenia afflicts, at one time or another, about one percent of the population, a fraction that is remarkably constant throughout the world and even across cultures. The disease unquestionably has a genetic component, even though the genes responsible have so far evaded some fairly intensive searches.

One strong piece of evidence for this genetic component comes from twin studies. In identical twins it has been repeatedly observed that if one twin is afflicted with schizophrenia, the other often develops the disease as well. The similarities may be even more striking, for sometimes the time of onset and the range and types of symptoms are very similar in both twins.

But not all affected pairs of identical twins show such a pattern. Sometimes only one twin develops schizophrenia and sometime they both do. The fraction of affected twins in which both develop the condition is called the concordance rate. If this always happens, the concordance rate will be a hundred percent. In fact, the rate for schizophrenia is only about half that. Half of the identical twin pairs are discordant for the disease.

This pattern is remarkably universal. A twin concordance rate well below a hundred percent is also seen in other major types of mental illness, although in each case the concordance rate is much higher than would be expected by chance. But if identical twins have identical genes, and genes are controlling this condition, shouldn't the concordance rates be a hundred percent? Why are there so many pairs of identical twins in which one twin develops a condition and the other does not?

The usual explanation given is that some traumatic event—an emotional crisis or a disease—must trigger the mental illness, and that about half the time it happens to one twin and not to the other. No candidates for such hypothetical events have yet been found, but a striking and little-noticed effect shown by these twins may provide a clue. Among discordant twin pairs, the twin who develops schizophrenia is almost always the lower birth-weight twin!

This remarkable observation, mentioned casually in a few papers on the subject, seems to me to hold great promise for understanding the causes of the disease. It may mean that schizophrenia can be triggered by something as simple as poor nutrition in utero. Or it may mean that the lower birth-weight twin is more susceptible to a disease of early childhood that brings on the mental illness, sometimes decades later. The genes give rise to the twig, but the environment must bend it.

If the reason for this birth-weight effect can be discovered, many cases of schizophrenia may be prevented simply by improving maternal nutrition during pregnancy. Or perhaps scientists will discover some disease or diseases that spur the condition and that preferentially affect the less well-nourished of the twin pair, perhaps shortly after birth. In sum, we would be well advised to investigate the roles of both early nutrition and childhood disease in the development of schizophrenia. The short-term payoff from these workaday studies may be far larger than the discovery of some of the many genes involved.

The interaction of schizophrenia and the environment raises a larger question: Do nutrition and disease have subtle effects on the mental capabilities of the rest of us? We will see in a moment that this may be so. The evolutionary consequences may be startling.

The bottom line of our search so far for the genes for behavior is that we are indeed beginning to discover genes that contribute, however subtly, to behavior patterns as we have narrowly defined them. But genes that contribute to behaviors in the more commonly understood sense continue to be as elusive as ever. We have found no genes that contribute to musical talent, to mathematical ability, or to other complex behaviors. We have certainly found no direct gene-culturgen relationships of the type suggested by Lumsden and Wilson.

We can now dismiss the paradox that there would have been no way to select for genes for piano-playing before the invention of the piano. This naïve view must be replaced by a much more subtle one, in which nature selects for complex sets of genes that interact in complex ways. An evolutionary process, involving many genes, has increased our brain capacity and flexibility over a broad front. This selection explains how we can turn our brains to a variety of tasks—like snowboarding—of which our forebears never dreamed.

Further, our brains have evolved into even better instruments than we suppose. It is often suggested, on the basis of rather specious arguments, that we use only ten percent of our brains. In fact, functional MRI scans show that we use a good deal more than that. But for a variety of environmental reasons, we do not do so very efficiently. Were we able to use more of our brain capacity, it would actually accelerate our evolution, because adaptive evolutionary change can take place only if physical or mental differences exist on which natural selection can act. If our environment reveals previously hidden genetic variation in our gene pool, then the effect of natural selection can become greater. The progressive unveiling of this variation, which is taking place as each succeeding human generation faces a more complex environment, can have unexpected results.

THE FLYNN EFFECT

Even those alien biologists from another planet whom we hypothesized in Chapter 10 would be forced to admit that the human brain is the most remarkable product of evolution to be found among the

Earth's living organisms. Although our culture has been shaped largely through the activities of people whose mental abilities are far above the ordinary, we cannot yet begin to define what those abilities are, or to understand what genetic differences might exist among us and how they have led to the occasional appearance of such outstanding people.

Despite the profundity of our ignorance, however, some interesting patterns do emerge. In the course of our evolution, our brains seem to have acquired capabilities that are more varied than even our current highly stimulating environment can elicit. The X-games of the future will involve feats of skill, balance, and coordination that we cannot begin to imagine, yet some of our children will be able to accomplish them with a flourish. Future societies will be more complex and challenging than any we have yet seen. Some of our descendants will thrive on the challenges, while others will be bewildered and discouraged. As we have already seen, such discouragement is likely to affect the gene pool, if people become so depressed and dysfunctional that they are unable to have children.

Our brains are capable of a wide variety of tasks. Sometimes they are damaged by biochemical and developmental hammers, padded or otherwise, but if they are not, they can manage very nicely. Such a basic or Model T brain, as we might call it, is perfectly capable of being tweaked, by training and by its inbuilt capabilities, in a great variety of directions.

All of us possess that basic brain. The various combinations of bells and whistles, the various factory options that each of us has in addition, are invisible even to the trained eye. Were we to line up the brains of an Einstein, a Mozart, and a Michael Jordan on a dissecting table, they would appear indistinguishable—mere "lumps of porridge," in the memorable phrase of psychologist Richard Gregory. While some differences must exist among them, my guess is that they are not large. The difference between a Mozart and an ordinary duffer who can just manage to pick out "Mary Had a Little Lamb" on the piano is a second-order difference.

One day soon we will be able to trace this particular difference to the number of synaptic connections between certain cells in the cortex, or to some unusual mix of neuropeptides. We will even be able to recreate the mix that ought to produce a Mozart and successfully apply it to any reasonably undamaged basic brain. Given such a biochemical and developmental boost and the right environment, any of us could then become a musical genius.*

We seem, though unconsciously, to be carrying out such manipulations already. In the early 1980s, James Flynn, a New Zealand psychologist, sent a questionnaire to researchers around the world. The researchers had been measuring intelligence test scores in their various countries over periods of years or decades. He asked a deceptively simple question: Did they see any trend? Was it true, for example, as is predicted by the prophets of genetic disaster, that IQ scores are going down because less intelligent people are having more children?

The period covered in the studies was from the 1950s to the 1980s. Flynn processed the masses of data with great care and reached a remarkable conclusion: During this time, rather than going down, IQ scores were in general going up, in some cases very far up. This increase appeared to be a very widespread phenomenon, since it was seen in all the industrialized countries of Europe, the United States, Australia, New Zealand, and Japan. The details of the pattern were puzzling: the largest gain was seen in the Netherlands, while one of the smallest was seen in nearby Great Britain.

For some subtests of the IQ tests, the average gain was as much as twenty percent. The gains were in many cases extremely significant and could not have been the result of some artifact of sampling. In the Netherlands, Norway, and France, which had universal conscription, IQ tests were given to virtually all males at age eighteen. Such data approach the statistician's ideal of measuring an entire population, rather than some sample that might or might not be randomly drawn.

The lengthy study that Flynn published in 1987 is festooned with caveats. First of all, the change can hardly have been an evolutionary one. Although evolution can be quick, it is not that quick. Probably some genetic alteration took place in the populations measured during those few brief decades, but it would primarily have been the result of mixing with immigrants—the very thing that racists predict should lead to a lowering of IQ. Flynn calculated that even the most extreme genetic scenario, in which genes play an overwhelming role in determining

---

* A tall order, probably, since the child would have to grow up with a group of people to whom music is absolutely central to existence and would in addition have to receive the undivided attention of the equivalent of a Leopold Mozart.

IQ, could have produced a change of only one percent or so each generation. Such a change, even if it were in the positive direction, could account for only a fraction of the huge increases. The effect must be due to changes in the environment.

Second, not all the components of the IQ tests showed the same increase in average score. Strangely enough, those that showed very little of the Flynn effect were the very ones that might have been expected to increase.

In recent decades succeeding cohorts of children have been exposed to more radio and television and more sophisticated schooling. One might suppose that they would therefore do better on so-called culture-loaded subtests, such as vocabulary tests that depend on special knowledge. In many cases there were increases in these subtests, but they tended to be small. Instead, the subtests that showed the largest increases were those that were perhaps the least culture-loaded.

One of these components is a test called Raven Progressive Matrices, in which the examinee must match ever more difficult patterns (Figure 12–1). While other parts of the tests are often rewritten to reflect changes in the tested population, this subtest has generally remained unaltered. It therefore serves as an excellent benchmark with which to follow a tested population over time. Flynn found that, on the whole, the young people of today are far better at solving these matching problems than their parents had been a generation earlier.

It is unlikely that these gains could have been due to better schooling. Indeed, by a variety of measures, schooling seems to have been getting less effective during the decades covered by the IQ measurements. Scores in widely applied tests of knowledge such as the American SAT and the English O- and A-levels have (until recently) been decreasing.

At the end of his paper, Flynn raised an interesting question. If one assumes that the IQ increases really do reflect an increase in intellectual ability, then why has the population at large not become measurably more brilliant with time? Surely far more geniuses should be abroad now than thirty years ago! As he looked about him, however, doubtless observing with horror the excesses and stupidities of the modern age, he failed to find evidence for such an increase.

Reading this part of his paper, one imagines that Flynn must have written it in a leather chair in his club, while harrumphing to fellow

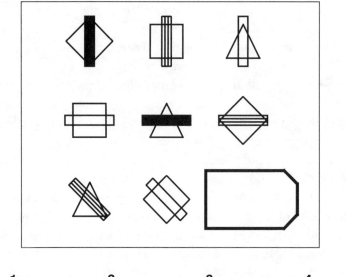

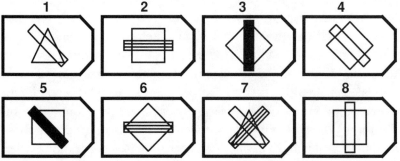

Figure 12–1. An example of Raven Progressive Matrices, not taken from an official test. The trick is to find the rules that make only one of the inserts below the figure the correct match. The answer is in the endnote for this page.

club members about the sad state of the younger generation. Nonetheless, he has a point: It is not terribly obvious that we are getting to be as brilliant as the numbers seem to indicate.

Flynn came to the tentative conclusion that the IQ subscores that he had been looking at so carefully may not have much to do with intelligence after all. Perhaps they were measuring some other factor, one that might not even be very strongly correlated with intelligence. It must be that other factor, he thought, that had changed so dramatically over the last few decades.

Perhaps he is right. But the tests do measure brain function at some level—and that function has changed in such a way as to improve our ability to do certain challenging tasks. Moreover, this effect has spilled over from the less culture-loaded to the more culture-loaded subtests, since smaller but significant gains were seen in a number of those that were more culture-loaded as well.

What this effect might be remains a mystery. Schooling, as we noted, seems to be ruled out. Three likely factors remain, all of which have the potential to influence overall brain function.

The first is television, a new factor that has become more important (and often central) to our lives during the postwar period. While it seems risible in the extreme to suppose that television as we know it might somehow be the cause of the Flynn effect, it is true that researchers at the University of Kansas have found that children who watch *Sesame Street* and its many imitators tend to perform better in school. The results of this study, unfortunately, are confounded by many other environmental factors that may be operating, notably class differences among the children. And *Sesame Street* is only a small part of the diet of television that most children watch.

But television may also have a more general effect, on speed of cognition. When people play fast-moving computer games, they quickly improve their cognitive skills. Perhaps the flashing of images in rapid succession during a television program or commercial can have a similar effect, speeding up reaction time and brain function. One certainly hopes so, and that the trillions of hours people spend in front of the tube worldwide every year are having some kind of benefit.

The second factor is nutrition. Since World War II diets in the industrialized world have improved markedly in terms of balance and completeness. During the last months of the war, the Dutch were deliberately starved by their German occupiers; after the war their diet became, to say the least, dramatically better. Flynn found that they show the largest IQ gains. On the other hand, the British, who showed the smallest overall IQ gains, were the last to benefit from this increase in the quality of nutrition. An insular nation, they have not yet completely abandoned the culinary habits that have made their cuisine a standing joke throughout the rest of the world. Having experienced a childhood of English food, I am quite willing to entertain the possibility that my intellect suffered as a result!

But the third factor, I suspect, is potentially the most important of all: the lessened influence of childhood disease. Measles, mumps, diphtheria, and whooping cough, so common only a few decades ago, may have had more of an effect on long-term brain function than we realize. Every child, as it grows up, suffers repeated bouts of illness, not all of which have an obvious cause. Some of these diseases, though slight in their immediate effects, may well leave behind a legacy of damage.

Perhaps not coincidentally, immunizations against these diseases have become more widespread and effective during the period of Flynn's study. Studies have shown that even children with moderately low birth weight suffer a higher incidence, and greater severity, of these childhood diseases than unimpaired children do. It is unclear whether the disabilities have caused the disease susceptibility or vice versa, but more detailed studies of these effects cry out to be done.

DISEASES AND THE MIND

The recent change in disease incidence has two evolutionary effects. The first is obvious: people whose brain function survives the on-slaught of disease are at an advantage. The second is that as the influence of disease lessens, these genetic differences are becoming less important to survival. Intellectual challenges from our environment, not disease-based challenges, will become more important in the future.

We do not know, however, how large a role diseases still play in brain function. If this role is still substantial, then we can examine which childhood diseases might have the greatest negative impact on brain function, for example, and why. Such studies will be important for the underdeveloped world, where many of these diseases are still prevalent.

In the industrialized world we have reduced or eliminated the most obvious diseases, but these are the easy targets. The ones that remain have less blatant effects, which will make their long-term harm-ful influences much more difficult to study.

That such early, almost invisible, damage can indeed have long-term consequences is becoming apparent. One of the most unnerving discoveries about brain function in recent years has emerged from an extensive study of Alzheimer's disease. About ten percent of elderly people, numbering perhaps four million in the United States, are affected by this debilitating and ultimately fatal neurodegenerative condition.

In 1986 epidemiologist David Snowdon of the University of Kentucky began to recruit nuns of the School Sisters of Notre Dame for a longitudinal study of the effects of aging. He soon found that this gigantic subject was too broad to study properly, given his limited resources, so he narrowed his focus and concentrated on the effects of one particular manifestation of aging, Alzheimer's disease.

The Sisterhood is a teaching order with convents scattered throughout the East and the Midwest. Snowdon succeeded in persuading 678 of the nuns to join the study. At the time of recruitment all his subjects were seventy-five or older. They provided information about themselves and their careers and agreed to undergo repeated psychological tests to follow the effects of aging. Most important, they also agreed to donate their brains for study after death.

The brains of the nuns who have died so far have been carefully preserved and examined under the microscope for the plaques and tangles that are symptomatic of Alzheimer's. Snowdon and his fellow researchers are now able to make direct correlations between symptoms of dementia observed before death and the types of brain damage that are due to this degenerative condition.

Many important results have already emerged from the study. A particularly striking one is that plaques and tangles in the brain do not necessarily result in dementia. Several of the nuns remained alert and intensely interested in the world around them right up until their deaths, and yet their brains showed signs of advanced Alzheimer's on autopsy. It will be very interesting to learn how their mental capabilities managed to survive a disease that, in its extreme form, literally eats tiny holes in the brain. Some of us are clearly far more resistant to the effects of Alzheimer's than others.

A second observation of great interest is that the onset of Alzheimer's can be sharply accelerated by other diseases, particularly strokes. Synergistic interactions between Alzheimer's and other disease states are only beginning to be explored. Infections of the brain with herpes virus, for example, may be associated with Alzheimer's, but this virus has also been found in normal aging brains. Its presence may simply mean that it has had time to spread into previously inaccessible tissues.

The third observation is perhaps the most remarkable of all. Half a century earlier, when the nuns were in their twenties and were being recruited into the order as young novitiates, many of them wrote short biographical essays. They made these essays available to Snowdon, on the chance that their early writings might contain some hint about their future fate.

Snowdon examined two characteristics of the essays: idea density and grammatical complexity. Idea density, as its name implies, is a measure of the content of the essay, and it can be quantified rather crudely as number of ideas per ten words. Grammatical complexity measures the number and type of sentences beyond the simple declarative, such as those that have dependent clauses or employ the conditional. Using these criteria, he divided the essays into the lowest-scoring third and the highest-scoring two-thirds and asked what proportion of each group belonged to the nuns whose brains showed signs of Alzheimer's.

The results for idea density seemed to be remarkably clear. Among the twenty-five now-deceased nuns who had written the essays, the brains of ten were found to show clear signs of Alzheimer's. Nine of these ten fell in the lowest third on the idea density scale. In contrast, among the remaining fifteen who showed few or no signs of the plaques and tangles, only two fell in the lowest third. This result was quite significant statistically.

At the same time, however, Snowdon found no significant pattern in the relation between the grammatical complexity of the essays and subsequent Alzheimer's. This finding was odd, because both idea density and grammatical complexity had been shown in previous studies to be good indicators of brain function. Why should the two results differ—unless the hypothetical early stages of Alzheimer's somehow affect only idea density and not grammatical complexity?

It is of course possible that the significant effect Snowden found was simply a statistical fluke, a function of the small sample for the study. This possibility, at least, should be resolvable. At the risk of sounding a bit ghoulish, time is on Snowdon's side. As the nuns inevitably die, the numbers in his database will increase—and so will his confidence in the results.

If we assume that idea density really is negatively related to the later development of Alzheimer's, this study seems to suggest one of two things.

First, Alzheimer's may be more likely to strike those who, for whatever reason, have been exposed to few ideas and whose prose reflects this. Indeed, a number of studies have shown that Alzheimer's is more prevalent among people of lower socioeconomic status. Perhaps this is because poor people are more likely to be poorly nourished, or to be exposed to diseases. These factors could serve as triggers for events that will eventually culminate in Alzheimer's.

Such people may simply be in the wrong place at the wrong time. If disease organisms are the ultimate villains in this train of events, then by the time Alzheimer's becomes apparent, the diseases will long since have disappeared. Such situations are not unknown for other diseases—rheumatoid arthritis, for example, can be triggered by a bacterial infection, causing life-threatening symptoms many years after the infection has been cured.

Alternatively, Alzheimer's may be built into the system, so that it starts wreaking its damage on the brain even while the fetus is still developing. Anybody, depending on the luck of the genetic draw, could then develop Alzheimer's. Reinforcing this idea is the growing body of evidence that the disease may have a genetic component; a number of genes have been found that are strongly implicated in the small minority of Alzheimer's cases that tend to run in families.

In fact, disentangling these genetic and environmental alternatives is almost impossible in practice. Probably, as with so many of these situations, we are seeing a contribution from both effects. If you are unlucky enough to draw the wrong set of alleles from the genetic lottery, but then are lucky enough to grow up in an environment free of the diseases that might trigger the condition, you might easily live to a ripe old age with all your faculties intact. But even in the absence of such alleles, a sufficiently severe combination of malnutrition and disease during childhood might be enough to trigger Alzheimer's. There are likely to be no simple answers.

Assuming the results of Snowdon's study hold up, they seem to fall into the pattern that we have seen again and again: Our brains are truly remarkable organs, but our environments are filled with factors that can prevent them from achieving their full potential. Some of these environmental insults are severe and obvious, while some are much less so.

Factors such as nutrition and childhood disease may lower the brain's capabilities in subtle ways, but they also may damage the brain so slightly that it still appears to be perfectly functional. Yet people who suffer this damage may lose the intellectual edge that they need to take full advantage of the opportunities provided to them by our current highly challenging environment.

Even in the face of damage, surprisingly minor environmental interventions can help bring the brain up to a higher level of function. One of the most important studies of the effectiveness of such intervention was a carefully randomized trial reported in the *Journal of the American Medical Association* in 1990. From eight different geographic regions around the United States, a total of almost a thousand children were chosen who had been born prematurely or otherwise had low birth weight. Numerous studies had already shown that such children were at high risk for mental retardation and behavior problems later in life. Could early intervention change this prospect?

The children were divided into two groups, carefully balanced for such factors as race, gender, and mother's educational level, but otherwise randomized. Both groups received normal follow-up pediatric care, but one group participated in a program that included home visits, periodic meetings with the parents, and intensive training. This training, certainly the most important component of the intervention, was carried out at Child Development Centers specially set up for the program. For two years, starting when they reached the age of twelve months, the children were bused to the centers and given intensive learning activities by teachers. There was one teacher for every three children.

The results were dramatic. In every geographical region highly significant differences in IQ score emerged between the two groups. The children in the intensive program scored an average of fourteen points higher (which is, by the way, the average difference in adult IQ scores between blacks and whites in the United States). The gain was a little less among the children who had been the most severely underweight when they were born, but it was still highly significant. The mothers of the children in the intervention group also reported fewer behavior problems.

It seems that such intervention really can have an effect. Similar results, though less well controlled, have been reported for Head Start, a U.S. government program that employs comparable though less intense intervention efforts. Head Start has primarily concentrated on preschool children, starting at age three, though it has now—in part as a result of the *JAMA* study—been extended to younger ages.

Unfortunately, follow-up studies of Head Start children have shown that the gains from this program are often lost when the children start regular schooling. Critics of the program have used such results to suggest that Head Start has no lasting effect and should be discontinued. (See that unpleasant book *The Bell Curve* for a sampling of these arguments.) I find this suggestion astonishingly stupid. In view of the fact that the gains from Head Start are clearly measurable, it would seem much more sensible to ask how regular schooling manages to undo them!

At least some of these factors are likely to be important in the Flynn effect, which leads me to be much more sanguine about his effect than Flynn is himself. To me, the Flynn effect says that we have recently begun to do a surprisingly good job of preventing damage to our brains and enhancing their capabilities. Over the space of a few decades, we have succeeded in ameliorating and perhaps even reversing much of the early-childhood damage that in earlier generations was almost unavoidable.

How we succeeded in doing this remains a mystery, since of course we do not yet know the nature of the damage. But whatever the mechanism of the Flynn effect may be, the effect itself has been strong enough to overcome the simultaneous onslaughts of family breakup, declining school standards, and the worst damage to our intellects that our popular culture can wreak. This is, let me emphasize again, not an evolutionary change but rather a change in the environment. However, as we will see, it can result in evolutionary change.

Taken together, all these studies lead to an important conclusion: *The brains of most of us are capable of working better than we suppose.* The genes that seem to be responsible for aberrant behavior patterns are now, on careful study, resolving themselves into crude genetic defects that bludgeon the brain biochemically or developmentally, and that are often enhanced by negative environmental factors such as disease and low birth weight.

I am not—repeat, not—suggesting that all human brains are equivalent. As a result of genetic differences, the abilities of human brains surely do differ from one to another. But as long as that basic brain of which I spoke earlier is not harmed by its environment or its owner, it is quite capable of doing an excellent job.

Once we realize this, the stigma attached to not being very bright will disappear, and blame for lower intelligence will be placed where it belongs: not with the individual but with deleterious factors in the environment.

Presumably our ability to enhance brain function through improvement in our diets, the prevention of disease, and alterations in the environment has limits. But if we deliberately set out to improve brain function through more direct chemical means or even through genetic manipulations, we can perhaps go much farther. Can it be that all of us are potential Mozarts or Einsteins, simply awaiting the appropriate biochemical manipulations to bring out these hidden talents?

Probably not—from their rarity, as I suggested earlier, such remarkable talents must result from highly fortuitous combinations of genes and environment. But one prediction seems safe: The mental capabilities of our grandchildren, while they may not be in the Einstein league, will be dramatically greater, on the average, than our own.

Demand for artificial enhancement of mental capacity is already great. "Smart" pills—combinations of plant products and other chemicals that are, one hopes, relatively innocuous—are currently available in health food stores. At the moment it is difficult to separate the placebo effect from anything that these pills might do to enhance brain function.

But a new generation of real "smart pills" that work on the structure and function of the brain cells themselves will soon be available. What would happen, for example, if that mysterious second protein involved in Williams syndrome were overexpressed in the dendrites of nerve cells, instead of being underexpressed as it is in people with the syndrome, or normally expressed as it is in most of us? Would it enable us to think in two and three dimensions with unparalleled ease? Or would it simply produce some new and damaging biochemical imbalance? We do not know, but we will probably find out soon, first in experimental animals and then—if it seems to work—inevitably in humans.

Now that we have a clearer idea of the potential resident in our brains, let us return to that deeper question: Why is that potential so great? Why can so many of us respond so effectively to the breathtaking new challenges presented by our environment, whether it be designing a Mars lander or learning how to snowboard? What genes have been, and are continuing to be, selected for?

GENETIC STUTTERS

In the course of our evolution, two things have happened to the genes that control the development of the human brain. First, the brain has become larger and more complex. In a straightforward evolutionary process, alleles of various regulatory genes that caused this enlargement must have spread through our population and become "fixed," replacing older alleles in every member of our species.

The outcome has been an increase in both the size and the capabilities of the basic or Model T brain that all of us now possess. It happened relatively quickly in evolutionary terms, building on genetic equipment that was already in place back when our brains were very much like those of chimpanzees.

Second, and more subtly, we have accumulated additional new alleles of many of these genes. We know this must have happened because the range of talents that we exhibit as a species has increased so markedly since our chimpanzeelike days, and at least some of this increase must be due to our genes. The addition of new alleles increases the diversity of a population, providing the bells and whistles or "factory options" that we talked about earlier. Because each gene may have many alleles, none of us can possess all this variation. We can have a maximum of only two allelic forms at each of our genes, regardless of how many different alleles there may be in the population at large. Other members of the population may have a quite different mix of alleles, which is one reason we are so diverse in our abilities.

Where are these alleles, and what are they like? We do not yet know, but one newly discovered category of genes provides an excellent place to look for them.

As we have seen, the relatively common type of mental retardation known as fragile X syndrome is caused by an expansion or lengthening of a short repeated bit of DNA on the X chromosome. In all of us this stretch consists of the three-base motif CGG, repeated over and over in a kind of genetic stutter. We carry it on our X chromosomes, and most of us have no more than twenty or thirty copies of the three-base motif in our repeated region.

Geneticists have given this and similar repeats various names, but one that is becoming increasingly popular is *microsatellite*. Satellite DNA is found in huge quantities in our chromosomes, consisting of quite long stretches of noncoding DNA repeated many times. It can be separated from the rest of the DNA in a centrifuge, where it forms a satellite band in the centrifuge tube. Microsatellites are far too small and far too thinly scattered around the chromosomes to be separated in this way, but the name has stuck.

Mutations are constantly appearing that, by adding or removing three-base motifs, make this bit of DNA longer or shorter. For the majority of us, it seems not to make much of a difference—this microsatellite is not even in a gene, although it happens to be very near a gene of unknown function named FMR1. But in people who have too many copies of the little motif—about a hundred or more—a runaway mutational process begins to happen. As the repeated region is passed down from one generation to the next it gets longer and longer—so long, indeed, that the two parts of the X chromosome that it connects begin to spread apart. The function of FMR1, and perhaps of many other genes nearby, is disrupted. The growing thread of out-of-control DNA connecting the two parts of the chromosome becomes so fragile that the chromosome often breaks just at this point.

Bad news indeed for the carriers of this out-of-control microsatellite. But this process is not limited to the fragile X syndrome: another microsatellite in another part of the X chromosome occasionally does the same thing, and other microsatellites on the rest of our chromosomes can also go out of control, breaking those chromosomes as well.

In addition many other, shorter microsatellite repeats are scattered throughout our chromosomes. They do not damage the chromosomes, but when they get to be more than forty or fifty motifs long they too tend to go out of control, getting longer and longer and giving rise to genetic diseases.

The striking thing is that virtually all these so-called repeat diseases affect brain function. More than a dozen of them are now known, and the list is lengthening swiftly. Some of them, like Huntington's disease, have received a great deal of publicity. Others, like myotonic dystrophy and Machado-Joseph disease, are much rarer and therefore less studied. But all have dramatic effects, and most of them result from having too many CGG or CAG repeats. Often these repeats are found inside genes, which means that the genes code for strange proteins that have long stretches made up the same amino acid—usually, but not always, glutamine.

When these distorted proteins are introduced into bacteria or into animal cells in culture, they can kill the cells, but nobody yet knows why. As you might imagine, hundreds of laboratories around the world are working on the problem.

It seems remarkable that so many of these odd and dangerous mutations affect brain function, rather than anything else. Why do some of them not affect liver function or the immune system? Perhaps so many of our genes are involved in brain function that repeat diseases are by chance more likely to affect the brain than something else. Alternatively, repeat mutations may indeed damage different organs, but because their consequences are less dramatic than the ones affecting the brain, they have not yet been discovered.

Even common mental illnesses like schizophrenia and manic depression may be influenced by these genetic stutters. One of the hallmarks of repeat diseases is that they can get worse as the distorted genes are passed down from one generation to the next and the repeat lengthens. This process has been given the name *genetic anticipation*— it is as if previous generations anticipate the more terrible consequences seen in generations yet to come. Although specific DNA repeats have yet to be traced to schizophrenia and manic depression, in some families these illnesses show signs of genetic anticipation. And recent studies have shown that unusually long repeats tend to be found among the genes of schizophrenics, though again the specific genes involved have yet to be tracked down.

Regardless of whether their association with brain function is real or a statistical accident, these microsatellites have intriguing properties. They can evolve very swiftly—and, as I have repeatedly emphasized, our large and highly flexible brains have been the result of swift evolution.

A decade ago (a lifetime in today's fast-moving scientific world) I wrote a book called *The Wisdom of the Genes* in which I predicted the existence of a class of genetic variants that can evolve swiftly—though I had no clear idea at the time of what they were like. This class, I suggested, would consist of genes that have become very good at evolving.

To make a convincing argument for such genes, I had to tread through a bit of a theoretical minefield. Obviously, neither organisms nor their genes can see into the future; genes cannot evolve quickly in order to protect some of us against an unexpected disaster, like the arrival of a giant asteroid. But over time genes can change swiftly and adapt their carriers to alterations in the environment that are more predictable—the coming and going of ice ages, the alternation of dry and wet periods in the climate, changes in the amount of sunshine that select for increases or decreases in skin pigment. If such environmental cycles are repeated often enough, they will select for genes that are capable of mutating quickly, adapting their carriers to these new but not totally unexpected environments.

I called such genes potential-generating genes, because they were capable of generating new evolutionary potential. Soon afterward one type of potential-generating gene was discovered, in an unexpected place, by several groups of workers, including Richard Moxon and his group at the University of Oxford. They found that many disease-causing bacteria carry short microsatellites, in or near a number of their genes. When these microsatellites change in length, they can turn the genes on or off.* The genes act as tiny switches, regulating the production of proteins that are found on the bacterial surface or of the toxins that the bacteria excrete.

Moxon, together with Richard Lenski of the University of Michigan, called these repeated regions contingency genes, because changes in their length could adapt the bacteria swiftly to different environmental contingencies. The striking thing was how quickly and easily these contingency genes could mutate back and forth, providing the bacteria with remarkable evolutionary flexibility.

All mutations can in theory be reversed by subsequent mutations that precisely undo the first change. But most of the time reverse mutations are rare—the chance that a given base change in a gene will be reversed is small, when you consider all the other base changes that are more likely to happen to the gene first. Microsatellites are highly unusual, because the mutation process is so reversible—they can become longer or shorter in units of one or more motifs, because the enzyme that copies the DNA is very prone to make such mistakes.

Having such contingency genes is very costly. If a bacterium must switch some genes off and others on in order to survive some new

---

* They do this by throwing the DNA of the genes in and out of reading frame. Because the information in a gene is read in groups of three bases, a repeat that consists of motifs four or five bases in length can throw the reading frame off if it becomes shorter or longer by one or more motifs. But subsequent mutations can easily put the gene back on track.

challenge from its host, then only the rare mutants that have this capability will survive. Most of the bacterial population will die. This is no problem for bacteria, which can quickly replenish their populations from a few survivors. But for more elaborate organisms such as ourselves it is a problem, which is almost certainly why such crude on-off contingency genes are not found in our own genomes.

We do, however, have many little microsatellites that are scattered among our own genes. Even though they are not serving as simple on-off switches, they may have more subtle functions. The burning question that our laboratory and many others are investigating is whether our little microsatellite stutters, too, can affect the genes that they are in or near. And if so, what might their effect be?

Moxon and I have speculated that they might be important in brain function, since so many of them, when they get out of control and become too long, can give rise to neurological conditions. Most of the time, luckily, they remain short and do not cause disease. But even in people without disease they are not all of one length—many alleles of different lengths are found at each of these potentially disease-causing microsatellites. These normal alleles do not cause disease, but do they have a function? Do they fine-tune our brains in some fashion, adding innumerable "factory options" to that basic brain?

Most of the microsatellites probably have little effect, but the list of microsatellites that have an influence is growing. The most striking example is a pair of microsatellites in a male sex hormone receptor gene. If one of the microsatellites is too long, it produces a neurological disease. But shorter alleles can have striking effects too—alleles that are not quite long enough to cause overt disease are associated with infertility, and very short alleles of both microsatellites are associated with prostate cancer.

Even if only a small fraction of the thousands of microsatellites in our genomes exert detectable effects, the number of possible combinations of options would be huge, because there are so many different-length microsatellite alleles in or near so many of our genes. Each of us is sure to have a different mix.

Since these alleles are known to mutate so swiftly, increasing and decreasing in length, the mix in our whole gene pool can change relatively quickly. Further, we know that these little stutters can come and go in the course of evolution. The microsatellites that we humans carry are sometimes in different places from the microsatellites found on the chromosomes of chimpanzees.

Exciting questions abound: Do humans have more of these little stutters than chimpanzees? If so, many new potential-generating mutations may have appeared in the course of our evolution. There are some indications, particularly from the laboratory of Simon Easteal at Australia's National University, that this may be so, although the data are difficult to interpret at the moment because most of the microsatellites have been found in humans and then looked for in chimpanzees, rather than the other way around. A concerted effort to find similar microsatellites unique to chimpanzees will provide a truer gauge of their prevalence in the two species.

Another question is whether neurological diseases caused by microsatellite repeat expansions are found uniquely in our own species or whether they can occur in chimpanzees and other apes as well. If they are unique to us, then perhaps they represent a genetic cost because our brains have evolved so quickly. Preliminary results suggest that this may be the case—several repeats that can lead to disease tend to have longer alleles in normal humans than in normal chimpanzees—though there is currently much argument over whether this, like Easteal's finding, may be a statistical artifact.

Schizophrenia, which afflicts one percent of the population, is a likely source of such genetic cost. Because schizophrenics tend to have few children, the genes involved should be removed rapidly from the population unless their damage is balanced by some benefit. Some early studies had suggested that the siblings of schizophrenics tend to be brighter than average, perhaps giving the genes a heterozygote advantage like sickle cell, but more careful work has now shown that the "unaffected" siblings of schizophrenics often show slight neurological deficits. This makes the prevalence of the disease even more puzzling.

As we saw earlier, there is growing evidence that repeat expansions may be involved in the disease. So, if repeat diseases are at work here, it is reasonable to wonder what possible advantage these repeats might have that counters the very large disadvantage of the disease. The answer may be hard to come by, since it may require a deeper understanding of how our brains work than we have at the present.

It is also important to ask whether the incidence of schizophrenia is as high in other primates as it is in us. If it is not, then the likelihood will be increased that this and other widely prevalent mental illnesses may be a cost incurred by our runaway brain evolution. But this, too, will be difficult to ascertain, because the symptoms of schizophrenia are likely to be very different in chimpanzees and gorillas. This will make the disease far more difficult to recognize in these primates— and goodness knows, schizophrenia is already hard enough to diagnose consistently in humans. Almost certainly, the answer will have to await the discovery of the genes that contribute to schizophrenia, followed by a comparison of these genes in ourselves and our close primate relatives.

HAVE WE DOMESTICATED OURSELVES?

Microsatellites can hardly be the whole explanation for the evolving diversity in the abilities of our brains. But they may help explain how we have been selected for greater and greater behavioral diversity in the past, and how we continue to be selected for even more diversity in the present.

If the growing diversity of our cultures is selecting us for genetic diversity, it is likely to be somewhat different from the diversity that our ancestors possessed. The emergence of new skills that were nec- essary for survival, and of continually shifting behavioral norms in our evolving societies, must have had an impact on our genes. How flex- ible are we, and how easily can we adapt to such shifting circumstances?

Neuroscientist Fred (Rusty) Gage of La Jolla's Salk Institute has recently reopened the old question of just how much our environ- ment really influences our brains. His recent experiments have built on a series of remarkable studies by previous investigators, chiefly a group at Berkeley who started work in the 1950s. In the Berkeley experiments rats were raised in either stimulating or deprived envi- ronments, and the effects on their brains were measured.

The researchers found that the brains of the stimulated rats had a thicker cortex, and that the connections between their neurons were more numerous and complex. But at the time it was not technically possible to determine whether the number of neurons had increased.

In the mid-1990s Gage carried out similar experiments with mice. He used a combination of techniques to show that very young mice that are exposed to a stimulating environment gain neurons as they grow up. The increase that he measured occurred in the dentate gyrus, a part of the temporal lobe that has been implicated in memory, though many other parts of the brain may have gained as well.

In addition, and remarkably, Gage and his group showed that the stimulated mice were cleverer than the others. Mice in both groups were put into a water "maze," essentially a round tank filled with opaque water. Like their enemies the cats, mice do not like to swim, but they will do so if necessary. A glass platform was placed at a specific point below the surface of the water, undetectable to the mice until they found it with their feet and were able to stop swimming.

After repeated trials, Gage and his colleagues found that the stimulated mice discovered the platform significantly more quickly than the unstimulated mice. Even though the stimulation they had undergone earlier had had nothing to do with water mazes, they did very well at meeting this new challenge. As a result of their environmental enrichment, their brains had become better at dealing with the unexpected.

His experiments also illustrated another aspect of evolution. His mice had, in a way, actually cooperated with the researchers in the course of the experiments because of their recent evolutionary history. This remarkable fact emerged when he discussed his experiments during a meeting of our human evolution group, and it illustrates vividly a central point of this book. The ways in which the mice had responded to the new challenge of the water maze were functions of both their environment and their evolutionary history.

The mice that Gage and his group normally used were laboratory animals, bred for docility. They swam about, found the platform, and then simply stood on it, relieved that they could finally stop swimming. This enabled the experimenter to pick them up and repeat the challenge.

But when mice of a strain that had recently been captured from the wild were used, the outcome was very different. The mice swam furiously about, found the platform—and then immediately leaped from it and scurried away across the laboratory floor!

These wild mice were particularly strong and active: perhaps the laboratory mice would have done the same thing had they been physically capable. But they didn't, and the maze experiment could therefore be used to measure the improvement in their intellectual capabilities.

Indeed, Gage's entire experiment had been designed to take advantage of the capabilities of laboratory, not wild, mice. He observed that when wild mice were put in the original enriched environment, they did not react the way the laboratory mice did. The laboratory mice rushed around and explored things, while the wild mice spent very little time exploring. As soon as they determined there was no food to be had, they tended to fall asleep in a corner.

As Gage soon found, measuring the brain power of the wild mice would have required a different sort of experiment. They had greater physical resources than the laboratory mice and had no hesitation about using them. Did they get smarter in their enriched environment? Perhaps not, since they seemed to pay so little attention to it.

In miniature Gage's experiment may illustrate what has happened during our own evolution. Like his domesticated mice, we can rise to new intellectual and even physical challenges, but we are probably doing so in a different way from the routes our ancestors took. Things might have been different for our species in the past, even the recent past, before people were tamed by civilization.

Suppose I could travel back in time to the court of Attila the Hun and attempt to give him an IQ test. He would probably have my head lopped off without thinking about it, and then get on with the business of subduing his enemies and raping their women.

We have, on the whole and for better or worse, intellectualized our world. Our efforts have made that world a tamer place and have modified our behaviors accordingly. Over time, in such a diverse but less dangerous world, selection for survival may have had effects on us that are similar to the selection for docility that had taken place in Gage's domesticated mice. Hun-like behavior is less advantageous than it once was. Perhaps we, too, have been tamed.

THIRTEEN

◈

The Final Objection

There exist no data which should lead a prudent man
to accept the hypothesis that IQ test scores are in any
degree heritable. This conclusion is so much at odds
with prevailing wisdom that it is necessary to ask, how
can so many psychologists believe the opposite?

LEON J. KAMIN, *The Science and Politics of IQ* (1974)

In the last few chapters I emphasized the roles that genes, and their interaction with the environment, have played in our rapid evolution. Most of my readers will, I assume, feel quite comfortable with a viewpoint that gives weight to both factors. But some will also be seething by this time. They will demand to know how I can suggest that genes play any role at all.

A number of researchers, notably Leon Kamin of Princeton University, have suggested that as far as they can see, no role for genes need be postulated to explain differences in the ways people think. Such genes may be there, they point out, but there is no evidence for it. Environmental variation can explain all the observed differences.

Kamin's arguments are interesting, and he has a superb ability to find flaws in the experimental designs of studies that psychologists and geneticists have carried out to measure the impact of genes on brain function. But flaws or not, one gigantic and unavoidable fact tells us that the genes must really be there.

This fact is simple: Our brains have evolved in the past. They could not have done so unless genes played a role. Evolution does not take place unless genetic differences exist on which natural selection can act.

As the Dutch botanist Wilhelm Johannsen first showed at the turn of the century, variation that is entirely the result of the environment cannot be selected for. He examined populations of highly inbred plants that had almost no genetic variation and found that the plants varied among themselves. The variation they displayed, however, was entirely due to environmental differences that were encountered by each seedling as it grew up. When he tried to select for seed size, he found that the seeds of subsequent generations were unchanged.

Critics of a role for genes will concede that humans have certainly evolved a great deal, both physically and mentally, in the past. But now, they will claim, everything has changed. We are no longer simply the prisoners of our genes, forced onto some continuing evolutionary path by factors beyond our conscious control.

Actually, our conscious control has probably played a huge role in our physical and mental evolution throughout the process. In *The Runaway Brain* I elaborated on the old idea that much of our rapid evolution over the last few million years has been the result of our response, not to changes in the natural environment, but to modifications that we ourselves have made in the environment. This progressive modification, and the ever greater and more human-oriented complexity that resulted, selected for those of our ancestors who were best able to take advantage of these new circumstances. The upshot was that we have been continually pushed further and further away from adaptation to the natural world.

This idea dates back, like so much else, at least to Darwin, who remarked in chapter 5 of *The Descent of Man* (1871):

> Now, if some man in a tribe, more sagacious than the
> others, invented a new snare or weapon, or other means
> of attack or defense, the plainest self-interest, without
> the assistance of much reasoning power, would prompt
> the other members of the tribe to imitate him; and all
> would thus profit. . . . If the new invention were an

important one, the tribe would increase in number, spread, and supplant other tribes. In a tribe thus rendered more numerous there would always be a rather better chance of the birth of other superior and inventive members. If such men left children to inherit their mental superiority, the chance of the birth of still more ingenious members would be somewhat better. . . . Even if they left no children, the tribe would still include their blood-relations; and it has been ascertained by agriculturists that by preserving and breeding from the family of an animal, which when slaughtered was found to be valuable, the desired character has been obtained.*

In my book I took this argument a step further and suggested that such events have escalated into a runaway process. Once this brain–body–environment feedback loop was established, it proceeded at an ever-increasing pace as the complexity of culture and the opportunity for new inventions increased. Because this feedback loop involves cultural changes, and these changes have accelerated dramatically, there is no reason to suppose that selection for increased intellectual capacity in our species has slackened.

Many genes can increase brain capacity in a general way, and these genes have been selected for. Additional variant alleles, accumulating in the population over a long span of time but not possessed by all of us, have provided us with a still more elaborate array of potential behavioral options. But, as we saw in Chapter 12, so many genes are involved that a given gene has no obvious connection to a given behavior. And the environment plays an enormous role.

This argument is rather different from the hereditarian–environmentalist dispute that has disfigured so much of social science—and that is responsible for so many of the conflicts in our society at large. Even extreme partisans of these two views know that the truth lies somewhere between, or at least they pay lip service to such a compromise. But it is useful to examine the assumptions that underlie these

---

* Here Darwin also anticipated the important evolutionary idea of kin selection, in which our inherited capacities can be passed on by our relatives even if we die without issue. One stands in awe of the man, who could cram two such seminal ideas into one paragraph!

extreme views in order to see what they imply in evolutionary terms. When we do so, the fundamental illogic of both extremes becomes glaringly apparent.

The extreme environmentalist view is that we all have the same innate capacities. Only because of unfortunate genetic or environmental accidents do some of us do less well in society than others. If this extreme were true, then selection *for* increased capabilities would be impossible, since there is simply be no genetic variation for such capabilities. Moreover, the environmentalist view implies that all our evolution has taken place in the past: For reasons that environmentalists have not explained, our brains are finished pieces of work, and we can only mess them up. The only natural selection that can operate today is selection against harmful mutations, such as the deletion that causes Williams syndrome. These mutations appear occasionally, damaging the brains of their carriers by bludgeoning them biochemically or developmentally.

The extreme hereditarian view, on the other hand, is that our capabilities are dictated entirely by our genotypes. Genetic variation is sufficient to explain all the differences among people. In this view we are all the prisoners of our genes; environmental manipulation will have no effect, or at least no lasting effect. An extreme hereditarian is able to predict confidently that the gains of Head Start will be only temporary, like the muscles produced by exercise that shrink away soon after we stop.

The particularly pernicious hereditarian view espoused by the authors of our favorite stalking horse, *The Bell Curve*, makes a further assumption: that all of us can be ranked from best to worst according to our genotypes. If our evolution is to continue, in this view, then we must all be sorted out quite mercilessly according to this ranking. After all, this is how it happened back in a state of nature, when only the best-adapted members of the population were able to survive and reproduce. But because we have been incautious enough to relax the rigors of natural selection, the less well-adapted genotypes have begun to survive, to the detriment of our species.

Partisans of both extremes agree that a relaxation of natural selection would be undesirable because it would allow deleterious genes to accumulate. To a pure environmentalist, however, most people carry a single good genotype, while bad genes are rare and distressing accidents.

To a pure hereditarian, most of us are bad, and only a few noble geno-types, especially those belonging to the hereditarians themselves, deserve to survive.

In the United States the seesaw battle between these extremes has coincided with the alternating political ascendancy of liberal and conservative views. The most recent triumph of the hereditarian view-point took place during that intellectual nadir known as the Reagan years. But this temporary dominance of the hereditarians is already beginning to melt away.

As we have seen, evidence is growing that crude genetic and environmental damage causes most cases of diminished intellectual capacity, and that without such damage anybody's brain is perfectly capable of functioning at a high level. Further, it has now been shown that early environmental intervention can have a dramatic effect. Even if the effect fades away in the absence of continued stimulation, this does not mean that the stimulation should not be carried out in the first place, merely that it must be continued throughout life. In these very important matters, the environmentalists are right.

The environmentalists are also winning on another front, where a bastion of the extreme hereditarian viewpoint has recently taken a severe battering. Over the years a number of apparently clear-cut stud-ies seemed to show that genes play a large role in determining how well we can think. The conclusions of these studies have now come into question.

The idea of using twins to study the influence of genes was first proposed by Darwin's cousin Francis Galton over a century ago. Since Galton's time, it has been realized that the best way to do so would be to find identical twins who were separated at birth and raised apart in different adoptive families. Identical twins have identical genotypes, but those pairs who are separated are subject to different environments, presumably as different as those encountered by two unrelated chil-dren. So if twins turn out to be very similar to each other, it must be because their genes have had a large influence, while if they are very different from each other, their different environments must have played the larger role.

It was very difficult to track down substantial numbers of sets of twins who really were separated shortly after birth and subsequently had nothing to do with each other. Moreover, the whole field of twin studies has suffered for years from severe embarrassment, after the discovery that one of its early practitioners, the late British psychologist Cyril Burt, faked an unknown but probably substantial part of his data. Burt was anxious to prove a large role for genes, and his political agenda seems to have overwhelmed his scruples. Nonetheless, a number of quite good data sets have recently been gathered, particularly by a group of psychologists headed by Tom Bouchard of the University of Minnesota. Their conclusions are remarkably similar to those of Burt, though the effect of genes that they find is not quite so strong.

In these data sets the IQ scores of the twins raised apart are found to be tightly correlated, almost as tightly as the scores of twins that have been raised together. These correlations can be used, after some statistical manipulation, to estimate the influence of the genes, and the results suggest that genes are responsible for at least sixty and perhaps seventy percent of the variation in IQ. This value, called heritability, is a rather complicated one, but taken at face value it suggests that genes play a very large role.

These results were used by the writers of *The Bell Curve* to claim that IQ meritocracies are being established in modern human societies. The children and grandchildren of people with high IQs will tend to have substantially higher-than-average IQs, and the reverse should also be true.

But a powerful argument has recently been made that the contribution of genes to IQ is surprisingly small, certainly smaller than the twin studies would suggest. According to this argument a factor that was downplayed in the earlier studies is much more important than anyone had supposed. That factor is the one part of their environment that even twins who were separated at birth share: the time they spent in the womb.

Bernie Devlin and his colleagues at the University of Pittsburgh School of Medicine and Carnegie Mellon University looked at the data from studies of identical twins raised apart. They also examined data from more than two hundred other studies of IQ correlations between various relatives, some raised together and some raised apart. To begin with, they noticed, as others had done before them, that twin studies by their very nature tend to emphasize the role of genes. The heritability estimates for twins raised apart tend to be much higher than estimates obtained from more distant relatives.

This is because the heritability measurements are different in the two situations. When twins are examined, all the effects of all their genes come into play, including the effects of those genes that depend for their expression on their genetic context—that is, on all the other genes that both twins happen to carry on their sets of chromosomes. However, when the twins marry different people and pass their genes on to the next generation, these genes are combined with the genes from the other parent and thus find themselves in a completely different genetic context. Some of these genes' effects will vanish or will alter in unexpected ways. Heritabilities will be lowered.

Thus, when the correlations between relatives who are more distant than twins are examined, lower heritabilities are found than those seen in twin studies. Only some of the influences of the genes that these relative share—those that can still be expressed in different contexts—can be detected.

This effect undoubtedly decreases the impact of the genes and makes the *Bell Curve* meritocracy idea less tenable. But Devlin and his colleagues discovered an additional reason for the greater heritabilities found in the twin studies.

Dizygotic or fraternal twins are the result of the fertilization of different eggs by different sperm in the same mother, and therefore they are no more genetically alike than siblings born at different times. But despite this resemblance, fraternal twins who have been raised together tend to show a much higher heritability for IQ than nontwin siblings who have been raised together.

This well-known observation had always been attributed to the fact that nontwin siblings, unlike dizygotic twins, are born at different times and as a consequence are raised under slightly different conditions. But Devlin and his group were struck by the fact that the effect was surprisingly substantial, and they wondered whether much of it could be attributed to the shared intrauterine environment of the fraternal twins before they were separated. If it could, then perhaps a lot of the resemblance between identical twins could also be attributed to their shared intrauterine environment.

Statistical analysis confirmed their hunch. When they built genetic models that ignored the effect of the intrauterine environment, and compared these models to all the data from the two hundred studies, the models fit the data very poorly. But models in which the intrauterine environment was taken into account fit the totality of the data very well. These models cut the heritability estimate in half, to a mere thirty-four percent. Thus only a third of the resemblance between IQs seen in relatives can be traced to their genes.

The environment inside the mother's womb is much more important in determining IQ than had previously been supposed. Further, it is astonishingly labile. Like a river that cannot be stepped into twice, a mother's intrauterine environment changes over time. It is this surprising fact that had misled earlier IQ workers—they had not imagined that the environment provided by an individual mother could change so much from pregnancy to pregnancy.

With their finding Devlin and his colleagues have effectively destroyed the IQ meritocracy idea. The genetic effect is so small that while parents and children may resemble each other for genetic reasons, the genetic connection between grandparents and grandchildren is far more tenuous. The genetic meritocracy predicted in *The Bell Curve* could never be established. It would be destroyed by both upward and downward mobility in very short order. There is no way of predicting who will be on the top of the meritocracy in the future.

Delvin and colleagues' findings should come as no surprise to readers of this book, for we have already seen how important the intrauterine environment can be. And that environment can vary even between fetuses carried in a single pregnancy: recall that the schizophrenic member of a discordant twin pair is almost always the lower birth-weight twin.

The fact that twins must share limited resources from their mothers can also work against both of them. Compared with their siblings, twins show a small but significant deficit in IQ, a deficit that various studies have shown to range from four to seven points. The intrauterine environment plays an immense role, and the more we learn about it, the greater we understand that role to be.

Environmentalists, however, need not celebrate a complete triumph. Genes do contribute substantially to IQ, so the evolution of our mental faculties has hardly come to a complete stop. The low heritability means that evolution might be a little slower, but it also means that it is taking place across a broader front.

This is perhaps the most exciting, and encouraging, conclusion that has emerged from the work of Devlin and his colleagues. No brilliant IQ meritocracy has forged ahead in the evolutionary race, leaving the rest of us in the intellectual dust. Instead, we are all playing a role, and a substantial role, in ongoing evolutionary change. The pool of genetic variation in our species is growing in diversity, both from new mutations and from the mixing together of disparate human groups. We are all threads in this grand evolutionary tapestry, and as our genes weave in and out and recombine from one generation to the next, hereditary meritocracies will play no part. We have been shaped, and continue to be shaped, not by a genetic elite but by a vast genetic democracy.

FOURTEEN

❖

Our Evolutionary Future

It would appear, then, that man has, physically
speaking, specialized in unspecialization and by this
means has won himself a new span of evolutionary life
and development. . . . He advances mentally or technically by modifying his environment to suit his needs
instead of, as heretofore, altering his physique and its
needs. . . . So he wins new extension in evolution, but
that change has henceforward to be with increasing
speed and increasingly psychological.

GERALD HEARD, *Pain, Sex and Time* (1939)

Mayflies survive as adults for only a few hours, before dying and falling in uncounted numbers back into the rivers and streams from which
they had so recently emerged. Humans, on the other hand, survive
for the biblical threescore years and ten, and these days most of us
manage to do even better. Still, in an evolutionary sense, both we and
mayflies exist for only a fleeting moment.

Unlike mayflies, humans think about our future, but we tend to
think only a few years ahead. The best prediction we can make about
the more distant future is to say rather feebly that it is likely to be
amazing. Consider our potential physical environment. The inhabitants of the Tibetan plateau have already penetrated the fringes of space,
but some of our descendants will go much farther. In a century or two

some of us will probably be living on other planets of our solar system and even on planets circling nearby stars.

Physiologist John West recently reminded our human evolution study group that the astronauts who visited the moon found themselves in a gravitational field one-sixth as strong as that of Earth. They quickly discovered that it was much more convenient for them to hop from place to place than it was to walk. How many generations would it take, on a low-gravity planet, for nature to select for the nimblest hoppers among the colonists? How long before their descendants' legs and pelvis bones changed as a result? Would their skeletons and musculature eventually converge with those of kangaroos? The rate at which such changes happen will depend on the severity of the conditions on the new planet and how necessary it will be to keep hopping in order to stay alive!

The biochemistries of the animals and plants that accompany the colonists will also change with time as they become adapted to new soil and light conditions. Inevitably the biochemistries of the colonists will follow suit. Eventually, perhaps within a few thousand years, the descendants of the colonists will have great difficulty surviving if they return to the earth. Because gene flow will be interrupted, they, like the animals and plants they have taken with them, will be on their way to becoming new species.

The new worlds themselves may be more numerous than we think. Some totally unexpected breakthrough in our understanding of physics may allow us to travel faster than light, or even to discover and colonize alternative universes. Such a breakthrough would greatly accelerate our expansion into many very different environments.

These individual colonists will doubtless be the adventurers among us, who have perhaps always been the ones to evolve most quickly. The adventurous ancestors of the people of the Gran Dolina and the Sima de los Huesos made their way to the farthest reaches of Europe from Africa and the Middle East more than a million years ago. Once isolated, these people went in a different direction from the ancestors of the rest of us and became the Neandertals.

In today's cramped, crowded world, such isolation is not possible, and indeed even the most isolated of living peoples are rapidly blending into the genetic mainstream of our species. But if, in the future, the region of the universe that we manage to inhabit becomes sufficiently large, such gene exchange will be impossible. Our species will once again fragment, as it has done so many times in the past. Science fiction writers have speculated about the possible consequences of such fragmentation. Our children's children are likely to live through the reality.

THE FUTURE OF OUR BRAINS

In the meantime, what about those of us who stay at home? We, too, will not be evolutionary couch potatoes. We will continue to change physically, probably at roughly the same rate that our species has always changed. But our biggest changes will undoubtedly come about because of our growing ability to explore the potentials of that basic human brain that I have been talking about in the last few chapters.

These capabilities of our brains have been building for millions of years. Harnessing them more fully will actually accelerate our evolution, though not necessarily in the directions that we might expect.

To understand this process, we must once again inquire into the various ways in which evolution works. In 1930 one of the great founders of the theory of evolutionary genetics, R. A. Fisher, made a simple but profound observation: The maximum rate of increase in evolutionary adaptation must be proportional to the amount of genetic variation for adaptability present in the population. This observation seems in retrospect obvious, and indeed Fisher was so certain that his insight had all the properties of a mathematical axiom that he named it the Fundamental Theorem of Natural Selection. The more variation there is, the more evolution can take place.

In the decades since, Fisher's theorem has been much argued over. For it to be true, the population's variation must not only be genetic, it must be selectable. Some of the variation detected in twin studies on IQ cannot be selected, as we have seen, because it depends so heavily on the interactions among various genes. As a result, it tends to disappear, or to change its effects dramatically, from one generation to another. This results in a paradox: a population may be full of genetic variation, but if some of this variation is so dependent on genetic and environmental context that it tends not to retain its character from one generation to the next, then natural selection will be unable to act on it.

A second restriction on Fisher's theorem is that a population may have a great deal of selectable genetic variation, but selection simply happens not to act on it in any systematic way. For instance, as we saw, chimpanzee populations are full of genetic variation, but they have changed very little for millions of years. Variation alone, even selectable variation, will not ensure that evolution proceeds in any particular direction. (It must be remembered that Fisher was dealing with the maximum possible rate of increase in fitness, not the rate that might actually be taking place in any given situation.)

A third restriction, and perhaps the most important, is that even if the genetic variation is both present and selectable, it simply may not matter in a particular environmental context. If some of the genetic variation has little connection to the appearance and capabilities of the organisms that carry it, then Mother Nature may as well save her breath to cool her porridge. No amount of selection will get anywhere.

Whether such genetic variation is revealed or not can depend strongly on what happens to the organism as it grows up. This was demonstrated decades ago in a series of remarkable experiments on fruit flies carried out by the embryologist C. H. Waddington.

Waddington began with the observation that fruit flies raised in the laboratory all tend to be very similar to each other; one would be hard put indeed to distinguish among them. But then he took eggs from those same flies and exposed them briefly to a severe environmental insult, a heat shock or brief exposure to ether vapor. The flies that resulted were much more variable in their appearance, and he was easily able to select for strains that had malformed wings or bodies. After only a few generations he could stop using the environmental shock on the eggs—his selected strains bred true even in the absence of the shock.

Waddington had at first supposed that any changes that he saw would be the result of the inheritance of characters acquired as a result of the stress. Such evolution is often called Lamarckian, although this label is not quite correct. Early in the nineteenth century the proto-evolutionist Jean-Baptiste Monet, the chevalier de Lamarck, had famously suggested that the long necks of giraffes have become stretched through their desire to reach tender leaves high in the trees. This lengthened neck was somehow passed down to subsequent generations.

Leaving aside Lamarck's rococo imaginings about desires influencing evolution, he presupposed an evolutionary mechanism in which characters acquired during the lifetime of an organism are somehow passed on to the next generation. This is what is generally meant by Lamarckian evolution. Waddington tried to hybridize Lamarck and genetics by suggesting that the effects of the developmental malformations he induced would somehow be "assimilated" into the gene pool of his flies.

The inheritance of acquired characters is a no-no among evolutionists, who now realize that what matters is changes in the genes, not in the bodies of the organisms that carry them. A sizzling response from Waddington's critics eventually put him right. He had not created the variation by the stresses he had used—the variation was already in the gene pool of the flies he started with. He had simply revealed it by shocking the eggs. The genetic variants were there all the time but did not disturb development except under unusual conditions.

What happens to fruit flies can happen to us. During the 1950s and 1960s, a ghastly and accidental Waddington-like experiment was carried out on our own species. The drug thalidomide, originally developed as a sedative, was also found to control morning sickness. Many pregnant women, particularly in Europe where it had received approval, took it. The dreadful consequences were soon apparent. Ten thousand babies were born who had limb deformities ranging from relatively mild to extremely severe.

Thalidomide has profound effects on fetal development, but not on all fetuses. Hundreds of thousands of pregnant women took the drug, but only a minority of them had deformed babies. Recent experiments on rats have shown that normal rats can be given high doses of thalidomide without detectable harm to their fetuses, while those that carry a mutant gene predisposing them to mild limb deformity prove to be much more sensitive. The human population must also have mutant alleles that, though normally unexpressed or only weakly expressed, will wreak havoc in the presence of thalidomide.*

The evidence is strong that we can reveal the effects of such genetic variation in our own species by drastic manipulations of the

* It is ironic that thalidomide is making a comeback—it turns out to be a very useful drug for treating leprosy and a wide variety of autoimmune diseases. But warnings about its dangers are going unheeded. A black market in the drug has sprung up, and thalidomide babies are already appearing again in Brazil, where leprosy is widespread.

environment. The thalidomide story is just one such case. As we saw in Chapter 2 with the Bhopal disaster, massive industrial accidents can reveal the effects of genes that would otherwise have remained concealed. But Bhopal is just the tip of the iceberg. Even milder environmental insults are likely to have profound genetic consequences, both to us and to the animal and plant species on which we depend.

The same process must take place under natural circumstances whenever the environment becomes more stressful. Evolution in any species, not just our own, is able to proceed more quickly under such stressful conditions, since more of the effects of the underlying genetic variation will be revealed.

But this process as it occurs in a state of nature differs profoundly from the experiments of Waddington. In the case of natural selection, the organisms that have the highest fitness in the stressful environment will be selected for, while those that suffer the most from the stress will be selected against. Waddington, who was acting as a clumsy stand-in for Mother Nature, did the opposite. He selected for, rather than against, the physical deformities that he could see under the microscope.

All this means a great deal for our future evolution. Just as physical stresses reveal genetic differences between individuals in their ability to develop normally, psychological stresses can reveal differences in individuals' ability to withstand such pressures, differences that might have had no impact in the absence of such pressures. Hierarchical differences among British civil servants affect their survival and perhaps their reproductive capability as well. Daily news reports tell us of the profound effects of psychological factors on public health around the world.

Such stresses need not be entirely harmful. Even though the environment in which most of us live is stressful psychologically, it undoubtedly is less physically dangerous than that of *Homo erectus* or even those of our parents and grandparents. But it is certainly much more complicated. What is not yet susceptible to precise measurement is how all this increased complexity is affecting our survival and our reproductive capabilities. But there seems no doubt that the environment of the late twentieth century is revealing more than ever of our underlying genetic variation for brain function and perhaps for other characteristics, and is accelerating our evolution as a result.

Our own direct efforts will soon be adding immeasurably to that environmental complexity. Forthcoming "smart" pills will extend the capabilities of our basic brain, enhancing the bells and whistles we all possess in various combinations, but they will only be the beginning. Once we understand basic brain function, nothing will stop us from providing our own bells and whistles. Hands may be wrung about the ethical questions that such biochemical and even genetic manipulations of brain function raise, but parents given an opportunity to produce smarter children will unquestionably take it.

Pills that alter physical characteristics are already widely used. Some parents of children who happen to be a little shorter than the average are demanding that they be dosed with human growth hormone. Physicians are complying in large numbers, and manufacturers are actually promoting this questionable use. Luckily such treatment is a little safer now that recombinant DNA technology is being used to make the hormone. The dozens of children who contracted Creutzfeldt-Jakob disease during the 1970s, from pituitary extracts of growth hormone obtained from cadavers, were early casualties of our apparently overwhelming desire for all our children to become as tall as Wilt the Stilt.

When true "smart" pills are unleashed on our species, it is difficult to imagine the consequences. The reactions to them will vary greatly, resulting in yet another Waddington experiment that reveals all sorts of hidden genetic variation. We cannot predict who will survive this new biochemical onslaught, or what the consequences will be for our gene pool, but the results will certainly be, to put it mildly, interesting.

Future modifications of our mental capacities need not be confined to what goes on inside our own skulls. Soon, probably in a matter of decades, we will be able to tie our brains directly into vast databases such as the Internet without the clumsy intermediary of computers and keyboards. Perhaps little voices will whisper both information and misinformation into our ears, or RoboCop-like screens of data will scroll continuously across our eyeballs.

Already, even without such unnerving aids, some of us can handle the Internet much better than others and sort out the sensible information from the deluge of pseudoinformation and garbage. My rather Luddite guess is that, in the future, the people who are unable to stop up their ears like the crew of Ulysses' ship are the ones who will fall victim to the siren call of the Internet. Most of them will not die, of course, except for those whose paranoia is fed to fatal extremes by unrestricted access to a crazed universe of conspiracy theories. But they will be effectively removed from the gene pool. The people who are able to stop up their ears are the ones who will still be able to interact with other human beings—and get on with the evolutionarily important business of having babies!

STRULDBRUGGS, INC.

The greatest modification to our bodies that is likely to take place over the next few decades will likely be a substantial lengthening of our life spans. As I write this, media excitement is raging about an enzyme called telomerase that modifies the ends of chromosomes known as telomeres. Normal body cells do not make telomerase, which means that with each cell division, their telomeres tend to grow shorter.* Eventually the cells senesce and die. Cancer cells, on the other hand, continue to make telomerase, and many of them can proliferate indefinitely.

When the telomerase gene is turned on in normal human or mouse cells, they go through more than the normal number of cell divisions before they senesce. Newspaper stories greeted this announcement as if the fountain of youth had been discovered, but in fact—as usual—the story is much more complicated. Mice that lack the enzyme altogether have survived for a number of generations in the laboratory, and their rates of aging and cancer seem unchanged. Telomerase, it appears, is only part of the story of aging.

Nonetheless, the mechanisms of aging will certainly be understood in far greater depth very soon. There is no obvious reason why we cannot lengthen our life spans—we already know that they can change and have done so over a relatively brief period of evolutionary time. Our maximum life span of 120 years is twice that of chimpanzees—the oldest chimpanzee for which we have records died at 57. Yet four million years ago the maximum life span of our ancestors must have been similar to that of chimpanzees. If our life spans have doubled over this relatively brief period, they could very well double again.

Just as with "smart" pills, life-span pills will reveal the underlying genetic heterogeneity of our species and will allow selection to

---

* Poet T. S. Eliot anticipated this when he remarked: "I grow old, I grow old—I wear the ends of my chromosomes rolled!"

act more effectively. The effect on our future evolution is likely to be both profound and unexpected.

A pill that could lengthen our life span without converting our bodies into a mass of cancer cells would be a very hot item. At the same time, it is hard to imagine a medication that would have a more destabilizing effect on society. Will the pill have very different effects on different people, lengthening the life span of some and shortening that of others? What will it do to our reproductive capabilities? As with thalidomide, malformed children might result. And the pill is likely to be expensive—will we condemn most of the developing world to short life spans while some of us, like Jonathan Swift's Struldbruggs, grow older and older and hoard the wealth of the planet? Such a situation is unlikely to persist for long: I envision two-hundred-year-old geezers hanging from lampposts while enraged mobs roam the streets.

Life span lengthening is just one of the new regions of phenotypic and genetic modification into which we are moving pell-mell. At a recent symposium on germline gene therapy held at UCLA, biotechnological hubris seems to have reigned supreme. Future technologies in which not just one but dozens or hundreds of genes could be changed simultaneously were touted. James Watson, the co-discoverer of the structure of DNA, insisted that any attempt to regulate such engineering would be "a complete disaster." Leroy Hood, one of the leaders of the Human Genome Project, pronounced: "We are using exactly the same kinds of technologies that evolution does."

Well, not exactly. The law of unintended consequences will ensure that most of our attempts to modify our bodies and our genes in the future will not work out in the way that we expect. Natural selection will render its cool and inexorable judgment on the hubristic excesses of biotechnologists.

In short, as we have done so often in the past, we are sure to go on providing many inadvertent opportunities for natural selection to act on us.

THE FUTURE OF DIVERSITY

The mixing of gene pools, along with continued mutation, means that our descendants will carry more alleles of the many different genes that influence behaviors. In millennia to come, as additional alleles are selected for, our species will continue to become more genetically diverse.

But scientists tend to feel very uncomfortable when confronted with such a genetically diverse population. What, they ask, is going to maintain all this diversity—why should it not be lost by chance or by other factors over time?

This is a real problem. The sickle cell allele, for example, will be maintained in the human population only as long as selective pressure from malaria is present. But sometime in the next century, and probably even sooner, we will conquer malaria. With that selective pressure gone and only its harmful effects left, the sickle cell allele will gradually disappear, as it is already starting to do in the United States. It will disappear even more quickly if we find an accurate and inexpensive way to fix the gene itself. So will all the other genes that depend for their continued existence on this disease—the thalassemias, G6PD deficiency, the elliptocytoses, and the rest of those junky mutations that we met in Chapter 4. The result will be a loss of population diversity, not a gain.

It might be argued that genes for behavioral diversity should be lost as well, since a uniform future awaits us. In science fiction futures, such as that portrayed in Fritz Lang's film *Metropolis*, huge hordes of people behave robotically because of the power exerted by a small elite. In the real world regimentation sometimes works too, at least temporarily. Early in this century Henry Ford regimented the assembly line, which helped to spark Lang's nightmare vision. But as you will remember, Charlie Chaplin managed to escape from that regimentation in *Modern Times*.

In 1968 I was aboard an Air Pakistan plane about to take off from Paris to Rome. At the last minute the plane suddenly filled up with members of a Chinese Communist delegation. They were all dressed in identical blue Mao suits and wore identical red and gold Mao buttons. They had identical stony expressions on their faces as they tried their best to ignore the blandishments of capitalism that were everywhere around them. It was almost impossible to determine the sexes of the delegates.

Such extreme regimentation, of course, has not lasted long, as any visitor to present-day Beijing or Shanghai will attest. And even older and far more rigid societal structures are being destroyed in the current tidal wave of change. The Indian caste system has managed to last for millennia, but it is now breaking down—the president of India is now an untouchable.

A few years ago I suggested a model for the maintenance of behavioral diversity, based on a remarkable phenomenon discovered in the 1960s by the geneticist Claudine Petit of the University of Paris. She found that mutant male fruit flies were more successful at mating with females if they were rare in the population than if they were common.

Female fruit flies are usually courted by a succession of males before they finally mate with one of them. Petit, and many others who later looked at the phenomenon, found that the rare males, regardless of their genotype, were somehow able to short-circuit this process. It would appear that the females become jaded by the repetitious blandishments of one similar male after another, but are more likely to respond to males that are different—regardless of what those differences might be.

This "rare male effect" is just one of many so-called frequency-dependent phenomena that operate in the natural world. Recently, I came across another and much more complex one. Collaborating with tropical ecologists Richard Condit and Stephen Hubbell, I found strong frequency-dependence in a Panamanian forest filled with hundreds of tree species. If a particular species of tree is sparse in one part of a forest and common in another, the sparsely scattered trees tend to reproduce themselves more rapidly than the common ones. We suspect that this may be the result of the action of pathogens that are specific to each tree species—the pathogens spread when the trees become common and cannot spread when they are rare.

Such frequency-dependence will help to maintain the diversity of tree species in the forest, just as the rare male effect should help maintain genetic diversity in a population of fruit flies. As I analyzed the forest data, I began to wonder: Might the phenomenon of frequency-dependence help maintain and even increase human behavioral diversity?

One fascinating and little-noticed aspect of frequency-dependent selection is that it reinforces itself. In a tropical forest, as tree species evolve in various directions and become more and more different from each other over time, they are less likely to be attacked by pathogens.

This is because, since each host species is becoming different from the others, most pathogens will be forced to specialize on particular host species. Given enough time, each tree species becomes so different from the others that there are no similar tree species with which they share pathogens. Thus diversity itself will be selected for. No wonder that tropical ecosystems, with their huge variety of hosts and pathogens, are brimming with diversity.

In our own species we may be seeing something similar, albeit at the behavioral level rather than at the level of host-pathogen interactions. The opportunities for human behavioral diversity are greater than they have ever been. As a result, more and more of our underlying genetic diversity is revealed. It is like a gigantic Waddington experiment, although the way in which this behavioral diversity is currently being revealed is considerably less draconian than the way in which Waddington revealed the underlying variation in his fruit flies.

If this uncovering of underlying genetic diversity for behavior is combined with frequency-dependent selection acting on behavioral characters, then we have a recipe for ever-increasing behavioral diversity.

In our society people with particular skills often do very well if those skills are unusual. There is, it seems, room in our world for only one Rupert Murdoch. If society were filled with many Rupert Murdochs, none of them could possibly be as successful as the original. While the genes that contribute to Rupert Murdoch–ness might be advantageous when they are rare, they would quickly become disadvantageous if they were to be common.

Of course, in the Darwinian sense, Rupert Murdoch can translate his advantage into an evolutionary one only if he or his close relatives have lots of children. There are no genes for acquiring newspapers and satellite broadcast systems, but the alleles that contribute to entrepreneurship in a more general sense, whatever they may be, are likely to be advantageous when they are rare and disadvantageous when they are common. Many other alleles that contribute, in a complex and highly interactive way, to a great variety of skills should show a similar frequency-dependence. Like that tropical forest, our species can be expected to continue to maintain and increase its diversity.

All this, it seems, is happening just at a time when we need it more than ever.

THE BOTTOM LINE

Our species is moving into a period of great but nonetheless manageable danger. Population pressure, while easing slightly, is still a terrible threat to our health and to the well-being of the ecosystems on which we depend. Nuclear, chemical, and biological warfare continue to be perils, particularly in the hands of nationalists and religious extremists trying to preserve intact their antiquated ideas and their already hopelessly contaminated gene pools. But these dangers, as serious as they are, are unlikely to threaten the survival of our entire species. Most of us live much safer lives than our grandparents did.

We have no guarantee that this safety will continue. A real ecological disaster may lie in our future. The ice caps may melt. The Gulf Stream may, as it has done repeatedly in the past, shift its path. A really devastating disease may suddenly appear, or an unexpected environmental contaminant may make us suddenly unable to reproduce. Our planet may at any moment collide with some rock from outer space. All these scenarios would reduce our population drastically. The survivors, like Dr. Strangelove hiding out with his secretarial pool in his cave, would be anything but a random sampling of our species. But the most probable scenario is that for the next few thousand years we will proceed as we have for the last few thousand, lurching from one near-disaster to the next even as most people manage to survive each generation.

Whether our physical environment becomes safer in the future or takes a turn for the worse, our intellectual environment is certainly becoming more challenging. It is in this realm that our future evolution will primarily take place, continuing and enhancing the trend that has continued uninterrupted for the last several million years. The challenges we will face—traveling to other stars, healing our damaged world, learning how to live with our differences—will be met in part because we will be able to draw on that genetic legacy.

We are the children of Promethean ancestors who set us on this remarkable evolutionary course. Fire was only one of the remarkable discoveries that they bequeathed to us. Are we still evolving? Because we must learn to deal with the costs of all those other Promethean discoveries, as well as with their benefits, it is very lucky for us that we are.

NOTES

INTRODUCTION

1 The epigraph quote from Patrick Synge is from Tom Harrisson, *Borneo Jungle: An Account of the Oxford Expedition to Sarawak* (London: Drummond, 1938).

3 Galdikas recounts her adventures in Birute Galdikas, *Reflections of Eden: My Years with the Orangutans of Borneo* (Boston: Little, Brown, 1995). Food use by the orangs is examined in R.A. Hamilton and B.M.F. Galdikas, "A Preliminary Study of Food Selection by the Orangutan in Relation to Plant Quality," *Primates* 35 (1994): 225–63.

5 The Ridley review is Mark Ridley, "Eco Homo: How the Human Being Emerged From the Cataclysmic History of the Earth," *New York Times Book Review* (17 August 1997): 11.

6 The Borneo fire story is told in Seth Mydans, "Southeast Asia Chokes as Indonesian Forests Burn," in *New York Times* (25 September 1997): A1. Orangutans were captured as they fled the fires, as reported by Mydans in, "In Asia's Vast Forest Fires, No Respite for Orangutans," in *New York Times* (16 December 1997): A1.

8 C.P. van Schaik, E.A. Fox, and A.F. Sitompul, "Manufacture and Use of Tools in Wild Sumatran Orangutans," *Naturwissenschaften* 83 (1996): 186–88, recounts a remarkable story of orang tool use. The orangutan project at the National Zoo is described in Bil Gilbert, "New Ideas in the Air at the National Zoo," *Smithsonian*

27 (1996): 32–41. Estimates of orang population size are made in Simon Husson, "On the Move: The Recent Discovery of Orangutans in the Peat Swamp Forests of Kalimantan Has Highlighted the Urgent Need to Protect This Fragile Ecosystem," *Geographic Magazine* 68 (1995): S2–S3.

9   Differences in Bornean and Sumatran orang chromosomes are reported in O.A. Ryder and L.G. Chemnick, "Chromosomal and Mitochondrial DNA Variation in Orangutans," *Journal of Heredity* 84 (1993): 405–409.

9   Efforts to keep the subspecies distinct are recounted in Lori Perkins, "AZA Species Survival Plan Profile: Orangutans," *Endangered Species Update* 13 (1996): 10–11. John Bonner, "Taiwan's Tragic Orang-utans," *New Scientist* 144 (1994): 10, gives some details of the illegal orang trade.

CHAPTER 1: AUTHORITIES DISAGREE

17–25   The quotations from this chapter come from: Ernst Mayr, *Animal Species and Evolution* (Cambridge, MA: Belknap Press of Harvard University Press, 1963); Jared Diamond, "The Great Leap Forward," *Discover* 10 (1989): 50–60; Richard G. Klein, "Archaeology of Modern Human Origins," *Evolutionary Anthropology* 1 (1992): 5–14; J.S. Jones, "Is Evolution Over? If We Can Be Sure About Anything, It's That Humanity Won't Become Superhuman," *New York Times* (22 September 1991): E17; J.V. Neel, "The Study of Natural Selection in Primitive and Civilized Human Populations," *Human Biology* 30 (1958): 43–72; Edward O. Wilson, *Sociobiology: The New Synthesis* (Cambridge, MA: Belknap Press of Harvard University Press, 1975); Roger Lewin, *Human Evolution: An Illustrated Introduction* (Boston: Blackwell Scientific Publications, 1993); David A. Hamburg, "Ancient Man in the Twentieth Century" in *The Quest for Man*, edited by V. Goodall (New York: Praeger, 1975); and C. Loring Brace, "Structural Reduction in Evolution," *American Naturalist* 97 (1963): 39–49.

19   Some of the papers detailing recent discoveries in pituitary growth hormone production in dwarfism mutants are W. Wu, J.D. Cogan, R.W. Pfaffle, et al., "Mutations in PROP1 Cause Familial Combined Pituitary Hormone Deficiency," *Nature Genetics* 18 (1998): 147–149; G. Baumann, and H. Maheshwari, "The Dwarfs of Sindh: Severe Growth Hormone (GH) Deficiency Caused by a Mutation in the GH-releasing Hormone Receptor Gene," *Acta Paediatrica*,

423 supp. (1997): 33–38; and J.S. Parks, M.E. Adess, and M.R. Brown, "Genes Regulating Hypothalamic and Pituitary Development," *Acta Paediatrica*, 423 supp. (1997): 28–32. An explanation for the short stature of African pygmies is set out in Y. Hattori, J.C. Vera, C.I. Rivas, et al., "Decreased Insulin-like Growth Factor I Receptor Expression and Function in Immortalized African Pygmy T Cells," *Journal of Clinical Endocrinology and Metabolism* 81 (1996): 2257–63.

19 Brace's estimates of the reduction in tooth size are in C. Loring Brace, "Biocultural Interaction and the Mechanism of Mosaic Evolution in the Emergence of 'modern' Morphology," *American Anthropologist* 97 (1995): 711–21.

20 Differences in wild and domestic cats are recounted in E. Fernandez, F. de Lope, and C. de la Cruz, "Cranial Morphology of Wild Cat (*Felis silvestris*) in South of Iberian Peninsula: Importance of Introgression by Domestic Cat (*Felis catus*)" (in French), *Mammalia* 56 (1992): 255–64. Charles Darwin emphasized the role of sexual selection in *The Descent of Man and Selection in Relation to Sex* (Princeton, NJ: Reprinted by Princeton University Press, 1981).

21 The history of eugenics is recounted in Daniel J. Kevles, *In the Name of Eugenics: Genetics and the Uses of Human Heredity* (Cambridge, MA: Harvard University Press, 1995).

26 The cartoon comes from page 106 of C. Loring Brace, *The Stages of Human Evolution: Human and Cultural Origins* (Englewood Cliffs, NJ: Prentice-Hall, 1967).

28 Surveys of rapid cichlid fish evolution are in L.S. Kaufman, L.J. Chapman, and C.A. Chapman, "Evolution in Fast Forward: Haplochromine Fishes of the Lake Victoria Region," *Endeavour* 21 (1997): 23–30. A recent book about the lake, Tijs Goldschmidt, *Darwin's Dreampond: Drama in Lake Victoria*, trans. Sherry Marx-Macdonald (Cambridge, MA: MIT Press, 1996), gives much fascinating information about this remarkable and fragile ecosystem.

34 The book I admired is Theodosius Dobzhansky, *Mankind Evolving: The Evolution of the Human Species* (New Haven, CT: Yale University Press, 1962).

CHAPTER 2: NATURAL SELECTION CAN BE SUBTLE

38 The prevalence of chromosomal abnormalities is documented in P.A. Jacobs, "The Role of Chromosome Abnormalities in Reproductive Failure," *Reproduction, Nutrition, Development*, Supp. 1

(1990): 63s–74s. The Bhopal disaster and new disasters in the making in India's "Golden Corridor" are recounted in Bruno Kenny, "Gujarat's Industrial Sacrifice Zones," *Multinational Monitor* 16 (1995): 18–21, while the effect of the disaster on pregnancies is examined in J.S. Bajaj, A. Misra, M. Rajalakshmi, et al., "Environmental Release of Chemicals and Reproductive Ecology," *Environmental Health Perspectives*, 101 supp. 2 (1993): 125–30.

38 Criticisms of the declining sperm count studies are summarized in Richard J. Sherins, "Are Semen Quality and Male Fertility Changing?" *New England Journal of Medicine* 332 (1995): 327. The Athens study that suggests there may be something to it is D.A. Adamopoulos, A. Pappa, S. Nicopoulou, et al., "Seminal Volume and Total Sperm Number Trends in Men Attending Subfertility Clinics in the Greater Athens Area During the Period 1977–1993," *Human Reproduction* 11 (1996): 1936–41.

40 Anderson's arguments about germ-line therapy are in John C. Fletcher and W. French Anderson, "Germ-line Therapy: A New Stage of Debate," *Law, Medicine and Health Care* 20 (1992): 26–39. R.E. Hammer, R.D. Palmiter, and R.L. Brinster, "Partial Correction of Murine Hereditary Growth Disorder by Germ-Line Incorporation of a New Gene," *Nature* 311 (1984): 65–67, recounts early transgenic mouse experiments with rat growth hormone. The more recent transgenic mouse experiments are found in J.E. Murphy, S. Zhou, K. Giese, et al., "Long-term Correction of Obesity and Diabetes in Genetically Obese Mice by a Single Intramuscular Injection of Recombinant Adeno-Associated Virus Encoding Mouse Leptin," *Proceedings of the National Academy of Sciences (U.S.)* 94 (1997): 13921–26.

41 The astounding results of these athlete surveys are reported in Michael Bamberger and Don Yaeger, "Over the Edge," *Sports Illustrated* 86 (1997): 60–67.

41 The long history of the domestic dog was quantified in C. Vilà, P. Savolainen, J.E. Maldonado, et al., "Multiple and Ancient Origins of the Domestic Dog," *Science* 276 (1997): 1687–89. The literature on the problems engendered by overselection in domesticated animals and plants is immense. One interesting and typical reference recounts experiments that show how outbred cattle outperform Herefords in many different areas: P.F. Arthur, M. Makarechian, R.T Berg, et al., "Longevity and Lifetime Productivity of Cows in a Purebred Hereford and Two Multibreed Synthetic Groups under Range Conditions," *Journal of Animal Science* 71 (1993): 1142–47.

43 The clever detective work that has gone into understanding the first stone tools is recounted in Glyn L. Isaac, *The Archaeology of Human Origins* (Cambridge: Cambridge University Press, 1989). B. Asfaw, Y. Beyene, G. Suwa, et al., "The Earliest Acheulian from Konso-Gardula," *Nature* 360 (1992): 732–35, reports the oldest Ethiopian stone tools, while the old Israeli artifacts and Pakistani tools were described by Avraham Ronen and Robin Dennell at the European Association of Archaeologists meeting, Ravenna, Italy, September 1997.

45 Some of the consequences of cranial synostosis are set out in P. Fehlow, "Craniosynostosis as a Risk Factor," *Child's Nervous System* 9 (1993): 325–27, and A.L. Albright, R.B. Towbin, and B.L. Shultz, "Long-term Outcome After Sagittal Synostosis Operations," *Pediatric Neurosurgery* 25 (1996): 78–82.

46 Details of the Gallic Roman tooth implant are found in E. Crubezy, P. Murail, L. Girard, et al., "False Teeth of the Roman World," *Nature* 391 (1998): 29.

47 The impressive study of the effects of ginkgo extract on Alzheimer's is P.L. LeBars, M.M. Katz, N. Berman, et al., "A Placebo-controlled, Double-Blind, Randomized Trial of an Extract of *Ginkgo Biloba* for Dementia," *Journal of the American Medical Association* 278 (1997): 1327–32.

47 A fruit fly memory-enhancing gene is reported in J.C. Yin, M. del Vecchio, H. Zhou, et al., "CREB as a Memory Modulator: Induced Expression of a dCREB2 Activator Isoform Enhances Long-term Memory in Drosophila," *Cell* 81 (1995): 107–15.

48–53 Ernest Beutler, a pioneer in the study of G6PD deficiency, reviews recent work in E. Beutler, "G6PD: Population Genetics and Clinical Manifestations," *Blood Reviews* 10 (1996): 45–52. The interaction between the genotype and smoking is reviewed in M.D. Evans and W.A. Pryor, "Cigarette Smoking, Emphysema, and Damage to Alpha 1-proteinase Inhibitor," *American Journal of Physiology* 266 (1994): L593–611.

48–49 Some aspects of the remarkably complex paraoxonase story are detailed in H.G. Davies, R.J. Richter, M. Keifer, et al., "The Effect of the Human Serum Paraoxonase Polymorphism is Reversed with Diazoxon, Soman and Sarin," *Nature Genetics* (1996): 334–36; Y. Yamasaki, K. Sakamoto, H. Watada, et al., "The Arg192 Isoform of Paraoxonase with Low Sarin-hydrolyzing Activity is Dominant in the Japanese," *Human Genetics* 101 (1997): 67–68; and M. Odawara,

Y. Tachi, and K. Yamashita, "Paraoxonase Polymorphism (Gln192-Arg) Is Associated with Coronary Heart Disease in Japanese Noninsulin-Dependent Diabetes Mellitus," *Journal of Clinical Endocrinology and Metabolism* 82 (1997): 2257–60.

51 An articulate expression of the concerns raised by environmentalists is found in Richard C. Lewontin, *Not in Our Genes: Biology, Ideology, and Human Nature* (New York: Pantheon Books, 1984). An utterly contrary view can be found in Roger Pearson, ed., *Shockley on Eugenics and Race: The Application of Science to the Solution of Human Problems*. (Washington, DC, Scott-Townsend Publishers, 1992).

51 The book that caused all the fuss is Richard J. Herrnstein and Charles Murray, *The Bell Curve: Intelligence and Class Structure in American Life* (New York: Free Press, 1994).

52 Many of Gardner's seminal ideas about intelligence are found in Howard Gardner, *Frames of Mind: The Theory of Multiple Intelligences* (New York: Basic Books, 1983).

CHAPTER 3: LIVING AT THE EDGE OF SPACE

55 The epigraph is from Tenzing Norgay, *Tiger of the Snows: The Autobiography of Tenzing of Everest* (New York: Putnam, 1955).

56–57 Something about these mysterious remains can be found in V.H. Mair, "Prehistoric Caucasoid Corpses of the Tarim-Basin," *Journal of Indo-European Studies* 23 (1995): 281–307.

58 General surveys of these adaptations are set out in two review papers: L.G. Moore, S. Zamudio, L. Curran-Everett, et al., "Genetic Adaptation to High Altitude" in *Sports and Exercise Medicine*, edited by S.C. Wood and R.C. Roach (New York: Marcel Dekker, 1994); and S. Niermeyer, S. Zamudio, and L.G. Moore, "The People" in *High Altitude Adaptation*, edited by R. Schoene and T. Hornbein (New York: Marcel Dekker, 1998).

58–59 The Caucasus site is reported in L. Gabunia and A. Vekua, "A Plio-Pleistocene Hominid from Dmanisi, East Georgia, Caucasus," *Nature* 373 (1995): 509–12. A readable popular account of the Yangtze find is in R. Larick and R.L. Ciochon, "The African Emergence and Early Asian Dispersals of the Genus *Homo*," *American Scientist* 84 (1996): 538–51.

59 Vadim A. Ranov, Eudald Carbonell, and Jose Pedro Rodriguez, "Kuldara: Earliest Human Occupation in Central Asia in its Afro-Asian Context," *Current Anthropology* 36 (1995): 337–46, and Susan

G. Keates, "On Earliest Human Occupation in Central Asia," *Current Anthropology* 37 (1996): 129–31, give contrasting views about the dating of the Yuanmou finds.

60–61 The complex history of Tibet is recounted in Hugh Edward Richardson, *Tibet and Its History* (London: Oxford University Press, 1962). Descriptions of the Paleolithic finds in the northern plateau can be found in Z. Sensui, "Uncovering Prehistoric Tibet," *China Reconstructs* 1 (1981): 64–65, and A. Zhimin, "Paleoliths and Microliths from Shenja and Shuanghu, Northern Tibet," *Current Anthropology* 23 (1982): 493–99.

62 The function of the Inca roads in the formation and dissolution of their empire is examined in Thomas Carl Patterson, *The Inca Empire: The Formation and Disintegration of a Pre-capitalist State* (New York: St. Martin's Press, 1991).

63 B. Arriaza, M. Allison, and E. Gerszten, "Maternal Mortality in Pre-Columbian Indians of Arica, Chile," *American Journal of Physical Anthropology* 77 (1988): 35–42, details the causes of death among the mummified Andean women. The *manta* cloth and how it works is explained in E.Z. Tronick, R.B. Thomas, and M. Daltabuit, "The Quechua Manta Pouch: A Caretaking Practice for Buffering the Peruvian Infant Against the Multiple Stressors of High Altitude," *Child Development* 65 (1994): 1005–13.

64–65 The remarkable resistance of Tibetan babies to the growth-slowing effects of high altitude is documented in Z.X. Zhoma, S.F. Sun, J.G. Zhang, et al., "Fetal Growth and Maternal Oxygen Supply in Tibetan and Han Residents of Lhasa," *FASEB Journal* 3 (1989): A987.

65–66 The reversal of the fetal reflex in pulmonary arteries of Tibetans was examined in B.M. Groves, T. Droma, J.R. Sutton, et al., "Minimal Hypoxic Pulmonary Hypertension in Normal Tibetans at 3,658 m.," *Journal of Applied Physiology* 74 (1993): 312–18. The striking ability of their babies to take up oxygen after birth is measured in S. Niermeyer, P. Yang, Shanmina, et al., "Arterial Oxygen Saturation in Tibetan and Han Infants Born in Lhasa," *New England Journal of Medicine* 333 (1995): 1248–52. Evidence for the putative gene for oxygen saturation can be found in C.M. Beall, J. Blangero, S. Williams-Blangero, et al., "Major Gene for Percent of Oxygen Saturation of Arterial Hemoglobin in Tibetan Highlanders," *American Journal of Physical Anthropology* 95 (1994): 271–76.

66 The Ladakh studies are in A.S. Wiley, "Neonatal Size and Infant Mortality at High Altitude in the Western Himalaya," *American*

Journal of Physical Anthropology 94 (1994): 289–305, while the comparative Sherpa and Quechua studies are in R.M. Winslow, K.W. Chapman, C.C. Gibson, et al., "Different Hematologic Responses to Hypoxia in Sherpas and Quechua Indians," *Journal of Applied Physiology* 66 (1989): 1561–69.

67 C. Monge, F. Leon-Velarde, and A. Arregui, "Increasing Prevalence of Excessive Erythrocytosis with Age Among Healthy High-Altitude Miners," *New England Journal of Medicine* 321 (1989): 1271, examines the puzzling prevalence of chronic mountain sickness at high altitude in Peru.

68 Breathing rates of Tibetans and Han are compared in J. Zhuang, T. Droma, S. Sun, et al., "Hypoxic Ventilatory Responsiveness in Tibetan Compared with Han Residents of 3,658 m.," *Journal of Applied Physiology* 74 (1993): 303–11. The puzzling lack of a similar high-altitude response among the Quechua of the Andes is recounted in R.B. Schoene, R.C. Roach, S. Lahiri, et al., "Increased Diffusion Capacity Maintains Arterial Saturation During Exercise in the Quechua Indians of Chilean Altiplano," *American Journal of Human Biology* 2 (1990): 663–68. The response of Sherpas and Quechua to artificial manipulations of oxygen level is examined in J.E. Holden, C.K. Stone, C.M. Clark, et al., "Enhanced Cardiac Metabolism of Plasma Glucose in High-Altitude Natives: Adaptation Against Chronic Hypoxia," *Journal of Applied Physiology* 79 (1995): 222–28.

71 Yak pulmonary circulation is compared with that of cattle in A.G. Durmowicz, S. Hofmeister, T.K. Kadyraliev, et al., "Functional and Structural Adaptation of the Yak Pulmonary Circulation to Residence at High Altitude," *Journal of Applied Physiology* 74 (1993): 2276–85. The remarkable history of the yak, and some features of its physiology and genetics, are recounted in Stanley J. Olsen, "Fossil Ancestry of the Yak, Its Cultural Significance and Domistication in Tibet," *Proceedings of the Academy of Natural Sciences of Philadelphia* 142 (1990): 73–100.

71 DNA was used to determine the relationship of yaks to other cattle in G.B. Hartl, R. Goeltenboth, M. Grillitsch, et al., "On the Biochemical Systematics of the Bovini," *Biochemical Systematics and Ecology* 16 (1988): 575–80.

72 The evidence for a possible allele advantageous at high altitudes is presented in H. E. Montgomery, R. Marshall, H. Hemingway, S. Myerson et al., "Human Gene for Physical Performance," *Nature* 393 (1998): 221.

CHAPTER 4: BESIEGED BY INVISIBLE ARMIES

74 The epigraph is from J.B.S. Haldane, "Disease and Evolution," *La Ricercha Scientifica*, 19 supp. (1949): 1–11. The details of how Linus Pauling became interested in sickle cell disease are taken from personal interviews with Harvey Itano and Jon Singer. A straightforward introduction to the history and symptoms of the disease is Miriam Bloom, *Understanding Sickle Cell Disease* (Jackson, MI: University Press of Mississippi, 1995).

77 The paper that resulted from this pathbreaking work is L. Pauling, H.A. Itano, S.J. Singer, et al., "Sickle Cell Anemia," a Molecular Disease," *Science* 110 (1949): 543–48.

77 Ingram's paper on the differences between normal and sickle cell hemoglobin is V.M. Ingram, "A Specific Chemical Difference Between the Globins of Normal Human and Sickle-cell Anemia Hemoglobin," *Nature* 178 (1956): 792–94.

79 The incidence study, one of several encouraging reports, is F.M. Gill, L.A. Sleeper, S.J. Weiner, et al., "Clinical Events in the First Decade in a Cohort of Infants with Sickle Cell Disease," *Blood* 86 (1995): 776–83.

80 Karen K. Kerle, Guy P. Runkle, and Barry J. Maron, "Sickle Cell Trait and Sudden Death in Athletes," *Journal of the American Medical Association* 276 (1996): 1472, reported on the American athlete case. The lack of effect of the allele on athletic career choice is documented in P. Thiriet, M.M. Lobe, I. Gweha, et al., "Prevalence of the Sickle Cell Trait in an Athletic West African Population," *Medicine and Science in Sports and Exercise* 23 (1991): 389–90.

83 Beet's earliest paper on sickle cell and malaria is E.A. Beet, "Sickle Cell Disease in the Balovak District of Northern Rhodesia," *East African Medical Journal* 23 (1946): 75–86.

83–84 The influence of blood cell rosettes on malaria is discussed in J. Carlson, G.B. Nash, V. Gabutti, et al., "Natural Protection Against Severe Plasmodium Falciparum Malaria Due to Impaired Rosette Formation," *Blood* 84 (1994): 3909–14.

85 Agriculture and its impact on the prevalence of disease is detailed in A.F. Fleming, "Agriculture-related Anaemias," *British Journal of Medical Science* 51 (1994): 345–57. The spread of the disease in Madagascar is told in S. Laventure, J. Mouchet, S. Blanchy, et al., "Rice: Source of Life and Death on the Plateaux Of Madagascar" (in French), *Santé* 6 (1996): 79–86.

86   The work of Livingstone and many others on this zoo of mutations is summarized in Frank B. Livingstone, *Frequencies of Hemoglobin Variants: Thalassemia, The Glucose-6-phosphate Dehydrogenase Deficiency, G6PD Variants, and Ovalocytosis in Human Populations* (New York: Oxford University Press, 1985).

89   The recent spread of these genes is examined in J. Flint, R.M. Harding, A.J. Boyce, et al., "The Population Genetics of the Haemoglobinopathies," *Baillieres Clinical Haematology* 6 (1993): 215–62. Ewald's views on disease evolution are set forth in Paul W. Ewald, *Evolution of Infectious Disease* (New York: Oxford University Press, 1994). I take a rather different view in Christopher Wills, *Yellow Fever, Black Goddess: The Coevolution of People and Plagues* (Reading, MA: Addison-Wesley, 1996).

90   The long independent evolutionary histories of *Plasmodium reichenowi* and *falciparum* are documented in A.A. Escalante and F.J. Ayala, "Phylogeny of the Malarial Genus Plasmodium, Derived from rRNA Gene Sequences," *Proceedings of the National Academy of Sciences (U.S.)* 91 (1994): 11373–77. Anopheline mosquitoes have difficulties in transmitting *falciparum* parasites, which are described in A.C. Gamage-Mendis, J. Rajakaruna, S. Weerasinghe, et al., "Infectivity of *Plasmodium vivax* and *P. falciparum* to *Anopheles tessellatus;* Relationship Between Oocyst and Sporozoite Development," *Transactions of the Royal Society of Tropical Medicine and Hygiene* 87 (1993): 3–6.

91   Koella's work on the effect of the parasite on mosquito survival is not yet published but is discussed in V. Morell, "How the Malaria Parasite Manipulates Its Hosts," *Science* 278 (1997): 223.

92   The impact of human diseases, particularly polio, on Goodall's chimpanzees is recounted in Jane Goodall, *In the Shadow of Man* (Boston: Houghton Mifflin, 1971).

92–93   V. Robert, T. Tchuinkam, B. Mulder, et al., "Effect of the Sickle Cell Trait Status of Gametocyte Carriers of *Plasmodium falciparum* on Infectivity to Anophelines," *American Journal of Tropical Medicine and Hygiene* 54 (1996): 111–13, measured the effect of being heterozygous for sickle cell on mosquito transmission. The original Red Queen paper was L. Van Valen, "A New Evolutionary Law," *Evolutionary Theory* 1 (1973): 1–30.

93–94   Cloning of the Duffy-negative gene is recounted in A. Chaudhuri, J. Polyakova, V. Zbrzezna, et al., "Cloning of Glycoprotein D cDNA, Which Encodes the Major Subunit of the Duffy Blood

Group System and the Receptor for the *Plasmodium vivax* Malaria Parasite," *Proceedings of the National Academy of Sciences (U.S.)* 90 (1993): 10793–97.

95   The story of how resistance genes to AIDS were discovered is told in Michael Dean and Stephen J. O'Brien, "In Search of AIDS-resistance Genes," *Scientific American* 277 (1997): 44–51.

96   The possible connection between plagues of the past and the HIV resistance allele is explored in J.C. Stephens, et al., "Dating the Origin of the CCR5- 32 AIDS Resistance Allele by the Coalescence of haplotypes," *American Journal of Human Genetics* 62 (1998): 1507–15. Gerard Lucotte, Serge Hazout, and Marc de Braekeller, "Complete Map of Cystic Fibrosis Mutation DF508 Frequencies in Western Europe and Correlation Between Mutation Frequencies and Incidence of Disease," *Human Biology* 67 (1995): 797–804, takes a detailed look at the CF gene in Europe.

97   The workings of cholera on nerve cells can be found in M. Jodal, "Neuronal Influence on Intestinal Transport," *Journal of Internal Medicine*, supp. 732 (1990): 125–32.

97–98   The mouse model for cholera was investigated by S.E. Gabriel, K.N. Brigman, B.H. Koller, et al., "Cystic Fibrosis Heterozygote Resistance to Cholera Toxin in the Cystic Fibrosis Mouse Model," *Science* 266 (1994): 107–109. Quinton's ingenious explanation for how CF mutants might have been selected was put forward in P.M. Quinton, "What Is Good About Cystic Fibrosis?," *Current Biology* 4 (1994): 742–43.

100   The dangers of *H. pylori* infections are recounted in G.C. Cook, "Gastroenterological Emergencies in the Tropics," *Baillieres Clinical Gastroenterology* 5 (1991): 861–86. Some of the many infectious agents associated with heart disease are discussed in R.W. Ellis, "Infection and Coronary Heart Disease," *Journal of Medical Microbiology* 46 (1997): 535–39. A recently discovered connection between a retrovirus and diabetes can be found in B. Conrad, R.N. Weissmahr, J. Boni, et al., "A Human Endogenous Retroviral Superantigen as Candidate Autoimmune Gene in Type I Diabetes," *Cell* 90 (1997): 303–13.

100–01   The confusing story about possible connections between infections and chronic fatigue syndrome is set out in C.J. Dickinson, "Chronic Fatigue Syndrome—Aetiological Aspects," *European Journal of Clinical Investigation* 27 (1997): 257–67.

CHAPTER 5: PERILS OF THE CIVIL SERVICE

104   The basic findings of the two Whitehall studies are set out in M.G. Marmot, M.J. Shipley, and G. Rose, "Inequalities in Death—Specific Explanations of a General Pattern?," *Lancet* 1 (1984): 1003–6, and M.G. Marmot, G.D. Smith, S. Stansfeld, et al., "Health Inequalities Among British Civil Servants: The Whitehall II Study," *Lancet* 337 (1991): 1387–93.

106   Studies that examine the effects of the fear of privatization are in J.E. Ferrie, M.J. Shipley, M.G. Marmot, et al., "Health Effects of Anticipation of Job Change and Non-employment: Longitudinal Data from the Whitehall II Study," *British Medical Journal* 311 (1995): 1264–69. The retirement follow-up is in M.G. Marmot and M.J. Shipley, "Do Socioeconomic Differences in Mortality Persist After Retirement? 25 Year Follow-up of Civil Servants from the First Whitehall Study," *British Medical Journal* 313 (1996): 1177–80.

107   The connection between heart disease and lack of control over one's life was explored in H. Bosma, M.G. Marmot, H. Hemingway, et al., "Low Job Control and Risk of Coronary Heart Disease in Whitehall II (Prospective Cohort) Study," *British Medical Journal* 314 (1997): 558–65.

108   Depression as a large predictive factor in heart diseases is explored in M.M. Dwight and A. Stoudemire, "Effects of Depressive Disorders on Coronary Artery Disease: A Review," *Harvard Review of Psychiatry* 5 (1997): 115–22, and A.H. Glassman and P.A. Shapiro, "Depression and the Course of Coronary Artery Disease," *American Journal of Psychiatry* 155 (1998): 4–11.

109   The connection between socioeconomic status and health in various societies is explored in N.E. Adler, W.T. Boyce, M.A. Chesney, et al., "Socioeconomic Inequalities in Health: No Easy Solution," *Journal of the American Medical Association* 269 (1993): 140–45. The remarkable mortality among unmarried people in Japan is documented in N. Goldman and Y. Hu, "Excess Mortality Among the Unmarried: A Case Study of Japan," *Social Science and Medicine* 36 (1993): 533–46.

111   Baboon hierarchies are explored in C.E. Virgin, Jr. and R.M. Sapolsky, "Styles of Male Social Behavior and Their Endocrine Correlates Among Low-Ranking Baboons," *American Journal of Primatology* 42 (1997): 25–39.

112   J. Altmann, R. Sapolsky, and P. Licht, "Baboon Fertility and Social Status," *Nature* 377 (1995): 688–90, and A. Pusey, J. Williams,
and J. Goodall, "The Influence of Dominance Rank on the Reproductive Success of Female Chimpanzees," *Science* 277 (1997):
828–31, present correlations of fitness with hierarchical positions
in baboons and chimpanzees.

113   Frans de Waal describes the remarkable behavior of monkey mothers in ensuring that their offspring associate with the "right" friends
in Frans de Waal, *Good Natured* (Cambridge, MA: Harvard University Press, 1996).

CHAPTER 6: FAREWELL TO THE MASTER RACE

114–15   Figures on the huge variation in human fertility from tribe to tribe
and region to region can be found in Lyliane Rosetta and C.G.N.
Mascie-Taylor, ed., *Variability in Human Fertility* (New York: Cam-
bridge University Press, 1996). Neel's computer model of the
Yanomama is described in J.W. MacCluer, J.V. Neel, and N.A.
Chagnon, "Demographic Structure of a Primitive Population: A
Simulation," *American Journal of Physical Anthropology* 35 (1975):
193–207. Chagnon's book about the Yanomama is Napoleon A.
Chagnon, *Yanomamo: The Fierce People*, 3rd ed. (New York: Holt,
Rinehart and Winston, 1983).

116   Estimates of the age of the tribal branch leading to the Yanomama
vary, but some careful calculations are given in R.H. Ward, H.
Gershowitz, M. Layrisse, et al., "The Genetic Structure of a Tribal
Population, the Yanomama Indians XI. Gene Frequencies for 10
Blood Groups and the ABH-Le Secretor Traits in the Yanomama
and Their Neighbors: The Uniqueness of the Tribe," *American
Journal of Human Genetics* 27 (1975): 1–30.

118   The huge gap between male and female life expectancies in Russia
is stabilizing but not shrinking—and many other societal pressures
are operating, as detailed in Bill Powell and Kim Palchikoff, "So-
ber, Rested and Ready: While Their Men Drown in Vodka and
Self-Pity, Russian Women Move On and Up," *Newsweek* (8 December 1997): 50–51.

118   Some of the truly horrifying effects of widespread pollution in Eastern Europe are documented in Irina Norska-Borowka, "Poland:
Environmental Pollution and Health in Katowice," *Lancet* 335
(1990): 1392–93.

118–20 Possible futures of the Russian population are charted in Carl Haub, "Population Change in the Former Soviet Republic," *Population Bulletin* 49 (1994): 2–47. Calhoun's famous experiments on the effects of overcrowding on rats are recounted in J.B. Calhoun, "Population Density and Social Pathology," *Scientific American* 206 (1962): 139–48. Freedman's studies on the effects of overcrowding in humans are in Jonathan L. Freedman, *Crowding and Behavior* (New York: Viking Press, 1975). Germany's population future is examined in Gerhard Heilig, Thomas Buttner, and Wolfgang Lutz, "Germany's Population: Turbulent Past, Uncertain Future," *Population Bulletin* 45 (1990): 1–47. The French survey of opinion on future population growth is reported in H. Bastide and A. Girard, "Les tendances démographiques en France et les attitudes de la population," *Population* 21 (1975): 9–50. Population transition in Korea is analyzed in Robert C. Repetto, *Economic Equality and Fertility in Developing Countries* (Baltimore: Johns Hopkins University Press, 1979), and in Taiwan in Ronald Freedman, Ming-Cheng Chang, and Te-Hsiung Sun, "Taiwan's Transition from High Fertility to Below-Replacement Levels," *Studies in Family Planning* 25 (1994): 317–31.

121 Predictions of the world's future population are examined in Barbara Crossette, "How to Fix A Crowded World: Add People," *New York Times* (2 November 1997): WK1. The prescient book written by Darwin's grandson is Charles Galton Darwin, *The Next Million Years* (New York: Doubleday, 1953).

121 Immigration and its impact on Germany are examined in Peter O'Brien, "Migration and Its Risks," *International Migration Review* 30 (1996): 1067–77. R.W. Johnson, "Whites in the New South Africa," *Dissent* 43 (1996): 134–37, is one of many studies on demographic trends in South Africa.

122 Japanese population history and current problems are vividly presented in Suzuki Kazue, "Women Rebuff the Call For More Babies," *Japan Quarterly* 42 (1995): 14–20.

123 An examination of the demographic revolution in California can be found in Dale Maharidge, *The Coming White Minority: California's Eruptions and America's Future* (New York: Times Books, 1996). Recent numbers on interracial marriages can be found in John Leland and Gregory Beals, "In Living Colors: Tiger Woods Is the Exception That Rules. For His Multiracial Generation, Hip Isn't Just Black and White," *Newsweek* (5 May 1997): 58–61.

CHAPTER 7: THE ROAD WE DID NOT TAKE

127   The epigraph is from Jared Diamond, "The Great Leap Forward," *Discover* (May 1989): 50–60.

127   Much information about Atapuerca and its excavations can be found in an issue of the *Journal of Human Evolution* that was devoted entirely to it: vol. 33, nos. 2–3 (1997).

130   The ancient child from the Gran Dolina is described in J.M. Bermudez de Castro, J.L. Arsuaga, E. Carbonell, et al., "A Hominid from the Lower Pleistocene of Atapuerca, Spain: Possible Ancestors to Neandertals and Modern Humans," *Science* 276 (1997): 1392–95.

132   Recent discoveries about *Homo heidelbergensis* are presented in M.B. Roberts, C.B. Stringer, and S.A. Parfitt, "A Hominid Tibia from Middle Pleistocene Sediments at Boxgrove, UK," *Nature* 369 (1994): 311–13.

132   An in-depth review of the meaning of Atapuerca can be found in R. Denell and W. Roebroeks, "The Earliest Colonization of Europe: The Short Chronology Revisited," *Antiquity* 70 (1996): 534–41.

132–33   The discoveries in the Sima de los Huesos are described in J.L. Arsuaga, I. Martinez, A. Garcia, et al., "Three New Human Skulls from the Sima de los Huesos Middle Pleistocene Site in Sierra de Atapuerca Spain," *Nature* 362 (1993): 534–37.

135   Differences in sizes between the sexes in the Sima de los Huesos are examined in J.L. Arsuaga, J.M. Carretero, C. Lorenzo, et al., "Size Variation in Middle Pleistocene Humans," *Science* 277 (1997): 1086–88.

138   The story of the German spears is told in H. Thieme, "Lower Palaeolithic Hunting Spears from Germany," *Nature* 385 (1997): 807–10.

139   Trinkaus's summary of the Neandertals is in Erik Trinkaus and Pat Shipman, *The Neandertals: Changing the Image of Mankind* (New York: Knopf, 1993). The remarkable Neandertal DNA story is described in M. Krings, A. Stone, R.W. Schmitz, et al., "Neandertal DNA Sequences and the Origin of Modern Humans," *Cell* 90 (1997): 19–30.

143   The final days of the Neandertals are described in J.J. Hublin, C.B. Ruiz, P.M. Lara, et al., "The Mousterian Site of Zafarraya (Andalucia, Spain): Dating and Implications on the Paleolithic

Peopling Processes of Western Europe," *Comptes Rendus de l'Academie des Sciences, Serie II: A Sciences de la Terre et des Planetes* 321 (1995): 931–37, and J.J. Hublin, "The First Europeans," *Archaeology* 49 (1996): 36–44. The Neandertal flower burials are described in Erik Trinkaus, *The Shanidar Neandertals* (New York: Academic Press, 1983).

143–44   Modified Neandertal stone tools are described in E. Boëda, J. Connan, D. Dessort, et al., "Bitumen as a Hafting Material on Middle Paleolithic Artefacts," *Nature* 380 (1996): 336–338. The flute is described in I. Turk, J. Dirjec, and B. Kavur, "Was the Oldest Music Instrument of Europe Found in Slovenia?" (in French), *Anthropologie* 101 (1997): 531–40.

144   Advanced artifacts at Arcy-sur-Cure and their association with Neandertals are detailed in J.J. Hublin, F. Spoor, M. Braun, et al., "A Late Neanderthal Associated with Upper Palaeolithic Artefacts," *Nature* 381 (1996): 224–26.

149   The predominance of genetic variation within rather than between human groups is measured by R.C. Lewontin, "The Apportionment of Human Diversity," *Evolutionary Biology* 6 (1972): 381–98. Mitochondrial population genetics of the Ladins is described in M. Stenico, L. Nigro, G. Bertorelle, et al., "High Mitochondrial Sequence Diversity in Linguistic Isolates of the Alps," *American Journal of Human Genetics* 59 (1996): 1363–75.

CHAPTER 8: WHY ARE WE SUCH EVOLUTIONARY SPEED DEMONS?

152   The discovery of our oldest ancestor is recounted in T.D. White, G. Suwa, and B. Asfaw, "*Australopithecus ramidus*, a New Species of Early Hominid from Aramis, Ethiopia," *Nature* 371 (1994): 306–12.

154   The story of the early Australopithecine discoveries is retold in Christopher Wills, *The Runaway Brain: The Evolution of Human Uniqueness* (New York: Basic Books, 1993).

154   M. Brunet, A. Beauvilain, Y. Coppens, et al., "*Australopithecus bahrelghazali*, a New Species of Early Hominid from Koro Toro Region, Chad," *Comptes Rendus de l'Academie des Sciences, Serie II: A Sciences de la Terre et des Planetes* 322 (1996): 907–13, is the source for the recent and surprising hominid discovery in West Africa.

155   The discovery of the First Family and its meaning is told in Donald C. Johanson, *Lucy: The Beginnings of Humankind* (New York:

Warner Books, 1982). The discovery of the fossil footprints can be found in Mary D. Leakey, *Disclosing the Past* (Garden City, NY: Doubleday, 1984).

156   A useful reference for events in human prehistory is S. Jones, R. Martin, and D. Pilbeam, eds. *The Cambridge Encyclopedia of Human Evolution* (New York: Cambridge University Press, 1992). The story of an ancient ape that seems to have had a remarkably upright stance is told in M. Koehler and S. Moya-Sola, "Ape-like or Hominid-like? The Positional Behavior of *Oreopithecus bambolii* Reconsidered," *Proceedings of the National Academy of Sciences (U.S.)* 94 (1997): 11747–51.

157   An excellent popular survey of the contributions of the Leakey clan to anthropology in Africa is Virginia Morell, *Ancestral Passions: The Leakey Family and the Quest for Humankind's Beginnings* (New York: Simon and Schuster, 1995). Susman's examination of the apparent tool-making capabilities of Australopithecine hands is R.L. Susman, "Fossil Evidence for Early Hominid Tool Use," *Science* 265 (1994): 1570–73.

159–60   Some of the confusion surrounding the skulls of *Homo habilis* is set out in G.P. Rightmire, "Variation Among Early *Homo* Crania from Olduvai Gorge and the Koobi Fora Region," *American Journal of Physical Anthropology* 90 (1993): 1–33.

161–62   Examination of Australopithecine brains has been carried out by R.L. Holloway, "Some Additional Morphological and Metrical Observations on *Pan* Brain Casts and Their Relevance to the Taung Endocast," *American Journal of Physical Anthropology* 77 (1988): 27–34, and P.V. Tobias, "The Brain of the First Hominids," in *Origins of the Human Brain*, edited by J.-P. Cagneux and J. Chavaillon (Oxford: Clarendon Press, 1995).

162   G. Philip Rightmire, *The Evolution of Homo erectus: Comparative Anatomical Studies of an Extinct Human Species* (Cambridge: Cambridge University Press, 1990), provides documentation for the increase in brain size of *Homo erectus* during its long tenure. The eventual end of the robust Australopithecines is told in B. Wood, C. Wood, and L. Konigsberg, "*Paranthropus boisei:* An Example of Evolutionary Stasis?," *American Journal of Physical Anthropology* 95 (1994): 117–36.

163–64   The story of the Georgia mandible is told in L. Gabunia and A. Vekua, "A Plio-Pleistocene hominid from Dmanisi, East Georgia,

Caucasus," *Nature* 373 (1995): 509–12. The history of the 'Ubeidiyah excavations is recounted in C. Guerin, O. Bar-Yosef, E. Debard, et al., "Archaeological and Palaeontological Programme on Older Pleistocene of 'Ubeidiya (Israel): Results 1992–1994" (in French), *Comptes Rendus de l'Academie des Sciences, Serie II A: Sciences de la Terre et des Planetes* 322 (1996): 709–12.

164–65   Much controversy surrounds the origin of the heterogeneous collection of archaic *Homo sapiens* remains. A small sampling of the conflicting points of view is set out in R.R. Sokal, N.L. Oden, J. Walker, et al., "Using Distance Matrices to Choose Between Competing Theories and an Application to the Origin of Modern Humans," *Journal of Human Evolution* 32 (1997): 501–22, and S. Sohn and M.H. Wolpoff, "Zuttiyeh Face: A View from the East," *American Journal of Physical Anthropology* 91 (1993): 325–47.

166   The possibility of early human-set fires in Australia is examined in A.P. Kershaw, "Pleistocene Vegetation of the Humid Regions of Northeastern Queensland," *Paleogeography Paleoclimatology Paleo-ecology* 109 (1994): 399–412.

167   The sad and complex story of Eugène Dubois and his discoveries can be found in Bert Theunissen, *Eugene Dubois and the Ape-man from Java: The History of the First Missing Link and Its Discoverer* (Boston: Kluwer Academic Publishers, 1989).

167–68   Swisher's oldest date for Javan *H. erectus* is in C.C. Swisher 3rd, G.H. Curtis, T. Jacob, et al., "Age of the Earliest Known Hominids in Java, Indonesia," *Science* 263 (1994): 1118–21, and his youngest is in C.C. Swisher 3rd, W.J. Rink, S.C. Anton, et al., "Latest *Homo erectus* of Java: Potential Contemporaneity with *Homo sapiens* in Southeast Asia," *Science* 274 (1996): 1870–1874.

170   Dating of Verhoeven's discovery was performed by M.J. Morwood, P.B. O'Suyllivan, A. Aziz, and A. Raza, "Fission-Track Ages of Stone Tools and Fossils on the East Indonesian Island of Flores," *Nature* 392 (1998): 173–76.

170   Rightmire's *Evolution of Homo erectus* gives the sizes of *H. erectus* brains. A good layperson's account of some of the Australian finds is in Josephine Flood, *Archaeology of the Dreamtime*, 2nd. ed. (Sydney: Collins Publishers Australia, 1989).

173   A brief survey of the recently emerging evidence for ancient cannibalism is Anne Gibbons, "Archaeologists Rediscover Cannibals," *Science* 277 (1997): 635–37.

CHAPTER 9: BOTTLENECKS AND SELECTIVE SWEEPS

174 C. Boesch, P. Marchesi, N. Marchesi, et al., "Is Nut Cracking in Wild Chimpanzees a Cultural Behaviour?," *Journal of Human Evolution* 26 (1994): 325–38, describes this remarkable behavior in detail.

176 Gagneux's discovery of naughty chimpanzee females is detailed in P. Gagneux, D.S. Woodruff, and C. Boesch, "Furtive Mating in Female Chimpanzees," *Nature* 387 (1997): 358–59.

178 Our mitochondrial DNA tree is found in P. Gagneux, C. Wills, U. Gerloff, et al., "Mitochondrial Sequences Show Diverse Histories of African Hominoids," *Proceedings of the National Academy of Sciences (U.S.)* (submitted). If you would like to construct such a tree, you can find out how in M. Saitou and M. Nei, "The Neighbor-joining Method: A New Method for Reconstructing Phylogenetic Trees," *Molecular Biology and Evolution* 4 (1987): 406–25.

181 Frans de Waal, *Bonobo: The Forgotten Ape* (Berkeley: University of California Press, 1997), is a magnficent book about the bonobos.

182 The checkered history of ape nomenclature is recounted in Colin P. Groves, "Systematics of the Great Apes," in *Comparative Primate Biology*, vol. 1: *Systematics and Anatomy*, edited by D. Swindler and J. Erwin (New York: Alan R. Liss, 1986).

183 The similarity between the variation in mitochondrial sequences of humans and of eastern chimpanzees was noted in T.L. Goldberg and M. Ruvolo, "The Geographic Apportionment of Mitochondrial Genetic Diversity in East African Chimpanzees, *Pan troglodytes schweinfurthii,*" *Molecular Biology and Evolution* 14 (1997) 976–84.

186 The mitochondrial Eve's name was first suggested by J. Wainscoat, "Human Evolution. Out of the Garden of Eden," *Nature* 325 (1987): 13. He was commenting on a seminal paper by the late Allan Wilson and his colleagues, R.L. Cann, M. Stoneking, and A.C. Wilson, "Mitochondrial DNA and Human Evolution," *Nature* 325 (1987): 31–36.

187 My own calculations about the mitochondrial Eve are in C. Wills, "When Did Eve Live? An Evolutionary Detective Story," *Evolution* 49 (1995): 593–607.

188 The hybridization paper is H. Vervaecke and L. Vanelsacker, "Hybrids between common chimpanzees (*Pan troglodytes*) and pygmy chimpanzees (*Pan paniscus*) in captivity," *Mammalia* 56 (1992): 667–69.

191 Our paper on the transitions and transversions is C. Wills, P. Gagneux, and S. Goldberg, "Selection Against Transversions in the Hominid Lineage," *Journal of Molecular Evolution* (submitted).

193 Doug Wallace's work on Leber's optic neuropathy and other mito-chondrial diseases is summarized in D.C. Wallace, "Mitochondrial DNA Mutations in Diseases of Energy Metabolism," *Journal of Bioenergetics and Biomembranes* 26 (1994): 241–50.

194 Hublin's work on the size of human and chimpanzee blood sup-plies to the brain is not yet published.

CHAPTER 10: STICKING OUT LIKE CYRANO'S NOSE

196–97 The genetic origins of the Ainu are examined in K. Omoto and N. Saitou, "Genetic Origins of the Japanese: A Partial Support for the Dual Structure Hypothesis," *American Journal of Physical Anthro-pology* 102 (1997): 437–46.

201–02 Loren C. Eiseley, *Darwin's Century: Evolution and the Men Who Dis-covered It* (New York: Doubleday, 1958), still provides an excellent examination of Wallace's ideas about human evolution. Osborn's ideas about the great antiquity of our species are set out in his in-troduction to Roy Chapman Andrews, *On the Trail of Ancient Man: A Narrative of the Field Work of the Central Asiatic Expeditions* (Gar-den City, NY: Garden City Publishing Company, 1926).

202 The molecular time scale for the branch point of the human and chimpanzee lineages is given in V.M. Sarich and A.C. Wilson, "Im-munological Time Scale for Hominid Evolution," *Science* 158 (1967): 1200–3.

206 Hundreds of papers have been written about genes found in hu-mans that are also present in chimpanzees, and occasionally vice versa. I know of no functional genes that are unique to one species or the other, although genes of a given type often differ in number.

207 A recent review of the extremely complex regulation of the mela-nin-producing enzyme tyrosinase can be found in C.A. Ferguson and S.H. Kidson, "The Regulation of Tyrosinase Gene Transcrip-tion," *Pigment Cell Research* 10 (1997): 127–38.

CHAPTER 11: GOING TO EXTREMES

211 Glen Dickson and Margot Suydam, "X Games Hit San Diego; ESPN Brings with It Virtual World, POVs and Fake Snow," *Broadcasting*

and Cable 127 (1997): 64–65, provides a brief description of the San Diego games.

214 The book about culturgens is C.J. Lumsden and Edward O. Wilson, *Genes, Minds and Culture* (Cambridge, MA: Harvard University Press, 1981). The rise and more recent decline of industrial melanism in *Biston betularia* is documented in B.S. Grant, D.F. Owen, and C.A. Clarke, "Parallel Rise and Fall of Melanic Peppered Moths in America and Britain," *Journal of Heredity* 87 (1996): 351–57.

216 The fate of normal children of retarded parents is examined in A.M. O'Neill, "Normal and Bright Children of Mentally Retarded Parents: The Huck Finn Syndrome," *Child Psychiatry and Human Development* 15 (1985): 255–68.

216 Pulmonary obstruction in Down's syndrome is detailed in S. Yamaki, H. Yasui, H. Kado, et al., "Pulmonary Vascular Disease and Operative Indications in Complete Atrioventricular Canal Defect in Early Infancy," *Journal of Thoracic and Cardiovascular Surgery* 106 (1993): 398–405. Recent findings about fragile X are summarized in A.T. Hoogeveen and B.A. Oostra, "The Fragile X Syndrome," *Journal of Inherited Metabolic Disease* 20 (1997): 139–51.

217 An excellent recent survey of the effects and genetics of Williams syndrome can be found in H.M. Lenhoff, P.P. Wang, F. Greenberg, et al., "Williams Syndrome and the Brain," *Scientific American* 277 (1997): 68–73.

218 The Williams gene story is in J.M. Frangiskakis, A.K. Ewart, C.A. Morris, et al., "LIM-kinase1 Hemizygosity Implicated in Imparied Visuospatial Constructive Cognition," *Cell* 86 (1996): 59–69.

219 Some of the many genetic deletions that lead to mental retardation are examined in M.L. Budarf and B.S. Emanuel, "Progress in the Autosomal Segmental Aneusomy Syndromes (SASs): Single or Multilocus Disorders?," *Human Molecular Genetics* 6 (1997): 1657–65.

CHAPTER 12: HOW BRAIN FUNCTION EVOLVES

221–22 The connection between genes and narcolepsy is traced in E. Mignot, "Perspectives in Narcolepsy Research and Therapy," *Current Opinion in Pulmonary Medicine* 2 (1996): 482–87.

223 A possible genetic basis for divorce is explored in Helen E. Fisher, *Anatomy of Love: The Natural History of Monogamy, Adultery, and Divorce* (New York: Norton, 1992). Some of the difficulties in

diagnosing ADHD are examined in S.C. Schneider and G. Tan, "Attention-Deficit Hyperactivity Disorder: In Pursuit of Diagnostic Accuracy," *Postgraduate Medicine* 101 (1997): 235–40.

224 P. Hauser, A.J. Zametkin, P. Martinez, et al., "Attention Deficit-Hyperactivity Disorder in People with Generalized Resistance to Thyroid Hormone," *New England Journal of Medicine* 328 (1993): 997–1001, was the first report on the thyroid hormone-ADHD connection.

224 The dopamine-serotonin connection with behavior patterns is explored in S.R. Pliszka, J.T. McCracken, and J.W. Maas, "Catecholamines in Attention-Deficit Hyperactivity Disorder: Current Perspectives," *Journal of the American Academy of Child and Adolescent Psychiatry* 35 (1996): 264–72.

226 Some of the difficulties and advances in tracking down genes for major mental illnesses are discussed in P. Sham, "Genetic Epidemiology," *British Medical Bulletin* 52 (1996): 408–33.

227 The connection between birth weight and schizophrenia, in both twins and in the general population, is explored in J.R. Stabenau and W. Pollin, "Heredity and Environment in Schizophrenia, Revisited," *Journal of Nervous and Mental Disease* 181 (1993): 290–97.

228 The disseminated nature of brain activity is a subject of much research. Some representative publications include H. Damasio, T.J. Grabowski, D. Tranel, et al., "A Neural Basis for Lexical Retrieval," *Nature* 380 (1996): 499–505, and M.I. Sereno, A.M. Dale, J.B. Reppas, et al., "Borders of Multiple Visual Areas in Humans Revealed by Functional Magnetic Resonance Imaging," *Science* 268 (1995): 889–93.

230–32 The Flynn effect, and the author's doubts about its meaning, are set out in detail in James R. Flynn, "Massive IQ Gains in 14 Nations: What IQ Tests Really Measure," *Psychological Bulletin* 101 (1987): 171–91.

231 The example of a Raven matrix is taken from Patricia A. Carpenter, Marcel A. Just, and Peter Shell, "What One Intelligence Test Measures: A Theoretical Account of the Processing in the Raven Progressive Matrices Test," *Psychological Review* 97 (1990): 404–31. The correct answer is 5.

233 Some possible benefits of watching educational television can be found in Aletha C. Huston and John C. Wright, "Educating Children with Television: The Forms of the Medium," in *Media,*

Children, and the Family: Social Scientific, Psychodynamic, and Clinical Perspectives, edited by D.J. Zillmann, J. Bryant, and A.C. Huston (Hillsdale, NJ: Lawrence Erlbaum Associates, 1994).

234   The severe effect of childhood diseases on low birth-weight children is documented in J.S. Read, J.D. Clemens, and M.A. Klebanoff, "Moderate Low Birth Weight and Infectious Disease Mortality During Infancy and Childhood," *American Journal of Epidemiology* 140 (1994): 721–33, and M.C. McCormick, J. Brooks-Gunn, K. Workman-Daniels, et al., "The Health and Developmental Status of Very Low-Birth-Weight Children at School Age," *Journal of the American Medical Association* 267 (1992): 2204–08.

235   Snowdon's study of an aging nun population is detailed in D.A. Snowdon, S.J. Kemper, J.A. Mortimer, et al., "Linguistic Ability in Early Life and Cognitive Function and Alzheimer's Disease in Late Life. Findings from the Nun Study," *Journal of the American Medical Association* 275 (1996): 528–32.

238   The *JAMA* study is "The Infant Health and Development Program. Enhancing the Outcomes of Low-Birth-Weight, Premature Infants. A Multisite, Randomized Trial," *Journal of the American Medical Association* 263 (1990): 3035–42.

238–39   The long-term effects of Head Start are examined by Constance Holden, "Head Start Enters Adulthood," *Science* 247 (1990): 1400–02.

242   Some of these triplet diseases are surveyed in R. Li and R.S. el-Mallakh, "Triplet Repeat Gene Sequences in Neuropsychiatric Diseases," *Harvard Review of Psychiatry* 5 (1997): 66–74. Some possible mechanisms by which glutamine repeats might cause cell death are discussed in C.A. Ross, M.W. Becher, V. Colomer, et al., "Huntington's Disease and Dentatorubral-Pallidoluysian Atrophy: Proteins, Pathogenesis and Pathology," *Brain Pathology* 7 (1997): 1003–16.

243   My book that discusses the evolution of evolvability itself is Christopher Wills, *The Wisdom of the Genes* (NY: Basic Books, 1989).

244   The paper suggesting the term *contingency genes* is E.R. Moxon, P.B. Rainey, M.A. Nowak, et al., "Adaptive Evolution of Highly Mutable Loci in Pathogenic Bacteria," *Current Biology* 4 (1994): 24–33.

245   Richard Moxon and I discuss possible reasons for microsatellite variation in E.R. Moxon and C. Wills, "Microsatellites: The Evolution of Evolvability" (tentative), *Scientific American* (in press).

246 Microsatellite effects are explored in T.G. Tut, F.J. Ghadessy, M.A. Trifiro, L. Pinsky, and E.L. Yong, "Long Polyglutamine Tracts in the Androgen Receptor are Associated with Reduced *Trans*-Activation, Impaired Sperm Production, and Male Infertility," *Journal of Clinical Endocrinology and Metabolism*, 82 (1997): 3777-782. The comparison between chimpanzees and humans is C.A. Wise, M. Sraml, D.C. Rubinsztein, et al., "Comparative Nuclear and Mitochondrial Genome Diversity in Humans and Chimpanzees," *Molecular Biology and Evolution* 14 (1997): 707–16. One of several studies finding differences in microsatellite disease gene repeats is P. Djian, J.M. Hancock, and H.S. Chana, "Codon Repeats in Genes Associated with Human Diseases: Fewer Repeats in the Genes of Nonhuman Primates and Nucleotide Substitutions Concentrated at the Sites of Reiteration," *Proceedings of the National Academy of Sciences* (US) 93 (1996): 417–21.

247 These early experiments measuring the effects of environmental enrichment on rats are detailed in Marian Diamond, *Enriching Heredity: The Impact of the Environment on the Anatomy of the Brain* (New York: Free Press, 1988). The recent experiments of Rusty Gage on mice are in G. Kempermann, H.G. Kuhn, and F.H. Gage, "More Hippocampal Neurons in Adult Mice Living in an Enriched Environment," *Nature* 386 (1997): 493–95.

CHAPTER 13: THE FINAL OBJECTION

250 The book that levels devastating criticism at a role for genes in IQ is Leon J. Kamin, *The Science and Politics of I.Q.* (New York: Halstead Press, 1974).

255 Bouchard's twin studies are in T.J. Bouchard, Jr., D.T. Lykken, M. McGue, et al., "Sources of Human Psychological Differences: The Minnesota Study of Twins Raised Apart," *Science* 250 (1990): 223–28. Devlin's important study is B. Devlin, M. Daniels, and K. Roeder, "The Heritability of IQ," *Nature* 388 (1997): 468–71.

CHAPTER 14: OUR EVOLUTIONARY FUTURE

259 The epigraph is from Gerald Heard, *Pain, Sex and Time: A New Outlook on Evolution and the Future of Man* (New York: Harper and Brothers, 1939).

261–62 Fisher's fundamental theorem is set out in R.A. Fisher, *The Genetical Theory of Natural Selection* (Oxford: Clarendon Press, 1930).

262–63   Waddington's first and mistaken idea appeared in C.H. Waddington, "The Canalization of Development and the Inheritance of Acquired Characters," *Nature* 150 (1942): 563. His important experiments appeared in C.H. Waddington, "Genetic Assimilation of an Acquired Character," *Evolution* 7 (1953): 1–13, and C.H. Waddington, "Genetic Assimilation of the *Bithorax* Phenotype, *Evolution* 10 (1956): 1–13.

265   The ghastly early history of treatment with human growth hormone is set out in S.D. Frasier, "The Not-so-good Old Days: Working with Pituitary Growth Hormone in North America, 1956 to 1985," *Journal of Pediatrics* 131 (Issue 1 pt. 2) (1997): S1–S4. The current ethical dilemmas are examined in Curtis A. Kin, "Coming Soon to the 'Genetic Supermarket' Near You," *Stanford Law Review* 48 (1996): 1573–1604.

265   Some of the media hype surrounding telomerase is described in Robert F. Service, "'Fountain of Youth' Lifts Biotech Stock," *Science* 279 (1998): 472. Experiments with mice lacking telomerase are described in M.A. Blasco, H.W. Lee, M.P. Hande, et al., "Telomere Shortening and Tumor Formation by Mouse Cells Lacking Telomerase RNA," *Cell* 91 (1997): 25–34.

267   A report on the UCLA meeting can be found in M. Wadham, "Germline Gene Therapy Must Be Spared Excessive Regulation," *Nature* 392 (1998): 317.

269   My behavioral diversity model is C. Wills, "The Maintenance of Behavioral Diversity in Human Societies" (comment), *Behavior and Brain Science* 17 (1994): 638–39. Petit's seminal discovery was first described in C. Petit, "Le déterminism génétique et psychophysiologique de la competition sexuelle chez *Drosophila melangaster,*" *Bulletin Biologique* 92 (1958): 248–329. Our rainforest paper is C. Wills, R. Condit, R. Foster, et al., "Strong Density- and Diversity-related Effects Help to Maintain Tree Species Diversity in a Neotropical Forest," *Proceedings of the National Academy of Sciences (U.S.)* 94 (1997): 1252–57.

GLOSSARY

※

ADHD (attention deficit hyperactivity disorder): This rather vaguely defined behavioral disorder usually involves short attention span, impulsive behavior, and high levels of physical activity. It has been connected with many different aspects of brain metabolism, in particular the effects of thyroid hormone and dopamine.

Ainu: An aboriginal people with rather Caucasian features but firmly in the East Asian gene pool, now confined to the island of Hokkaido in northern Japan.

Allele: An alternative form of a gene. The ABO blood group alleles are good examples, but there are many others. No individual can have more than two alleles of a gene, but a population may have many different alleles. Alleles are shuffled each generation by recombination.

Alzheimer's: A neurodegenerative disease, usually developing in middle or old age but occasionally sooner, in which the brain becomes filled with bits of precipitated protein and other structures forming plaques and tangles. Symptoms include short-term memory deficit and behavioral changes.

Anopheles mosquitos: The chief insect vectors of human malarias.

Aurignacian: Of or relating to late old stone age (Paleolithic) culture. Aurignacian tools, used by the Cro-Magnons and other invaders of Europe, are patiently shaped blades and scrapers, often with a razorlike edge and an elegant appearance. They are far more advanced than the crude stone tools of the Oldowan, but far less advanced than the beautiful stone and bone tools of the Neolithic.

Australopithecines: A heterogeneous group of hominids who lived throughout Africa from more than three to about a million years ago. Some had large teeth and huge jaws, and some were rather less robust. Their brains were roughly the same size as those of chimpanzees, but they may have been able to use tools.

Balanced polymorphism: A situation in which a balance of opposing forces maintains more than one allele in a population. The classic case is sickle cell anemia, in which the heterozygote is fitter than either homozygote. But a balanced polymorphism may arise in other ways. For example, if an allele is advantageous when it is rare but disadvantageous when it is common, this frequency-dependent selection can lead to a balance.

Base: One of the smaller molecular components of DNA. The four types of bases are called, in biological shorthand, A, T, G, and C. Like letters in a word, or words in a sentence, the order in which the bases are arranged along the DNA molecule encodes its genetic information.

Bonobo: A chimpanzeelike primate, found south of the Zaire River in Central Africa. The ancestors of bonobos and chimpanzees probably diverged about three million years ago.

Bottleneck: A dramatic reduction in a population's size, followed by an increase. The result may be a loss of genetic variability.

Chatelperronian: Stone tools and other artifacts used by the Neandertals thirty to forty thousand years ago. They have some characteristics in common with tools used by Cro-Magnons and other later invaders of Europe but are less sophisticated.

Chromosome: A structure in the nucleus of a cell that contains DNA. Mitochondria have chromosomes too, but they are much smaller and simpler than the chromosomes in nuclei and carry far fewer genes.

Codon: A group of three bases making up one unit of the DNA genetic code. Each codon specifies either an amino acid in the protein or a stopping point for the protein.

Contingency gene: A class of microsatellite, found in bacteria, that allows certain important bacterial genes to be switched on or off. These genes, with their high mutation rates, enable bacteria to adapt readily to new environmental contingencies; hence their name.

Cro-Magnon: An Upper Paleolithic group of *Homo sapiens* that migrated into western Europe, starting probably about 35,000 years ago. Cro-Magnons either drove the Neandertals to extinction or genetically assimilated them.

Cystic fibrosis: A genetic disease, common among Europeans, caused by a disturbance in the transport of chloride ions across the membranes that surround cells. This apparently minor defect can have fatal consequences. CF is thought to be maintained in the population by a balanced polymorphism.

Cytoplasm: The region of a living cell that surrounds the nucleus, where protein manufacture and metabolic activity are carried out. Mitochondria are found in the cytoplasm.

Divergence time: The time at which two evolutionary lineages diverged. Neither the fossil record nor the molecular record yet allows us to date such times with precision. Very large errors are associated with the times given in this book and elsewhere.

DNA (deoxyribonucleic acid): The genetic material carried by most living organisms. These long molecules, which form a double helix, comprise genetic information coded in a sequence of bases, like letters in a sentence.

Down's syndrome: A condition marked by mental retardation, heart abnormalities, and many other problems, caused by the presence of three copies of the small chromosome 21 rather than the normal two.

Duffy-negative: An allele that, when homozygous, essentially protects the carrier against malaria caused by *Plasmodium vivax*, because the parasite cannot penetrate their red blood cells. The allele is very common in central Africa.

Elliptocytosis: A condition that changes the shape of red blood cells and protects against malaria. It is particularly common in New Guinea.

Evolutionary tree: A branching diagram that arranges the physical features of organisms, or sequences of genes from organisms, so that the most closely related are nearest to each other and the most distantly related are farther away. Even the best of these trees involve some uncertainty, since the characteristics of organisms that lived in the past must be inferred from those that are living at the present time.

***Falciparum* malaria:** The severest form of human malaria, which can sometimes invade the brain. It is now largely confined to the tropics.

Flynn effect: The inexplicable rise in IQ scores in the developed world over the last few decades.

Fragile X: One of the best-known so-called repeat diseases, in which a small, highly repeated part of the X chromosome (a microsatellite) becomes very long and damages the ability of nearby genes to function.

Fraternal twins: Twins that result from multiple ovulations, in which two eggs are fertilized by different sperm. They are genetically as alike as brothers and sisters born at different times, but their shared intrauterine environment may give them more similar phenotypes. (*See* identical twins.)

G6PD deficiency: A very widespread genetic condition, common in the Mediterranean and Africa, that protects against malaria. Carriers of the gene, however, are very sensitive to oxidizing agents such as those found in broad beans.

Gene flow: An exchange of genes between two or more different populations. If the populations differ in allele frequencies, the result can be rapid genetic change.

Gene pool: The entire collection of genes in a population.

Gene: A stretch of DNA that carries some sort of genetic information. Many genes code for proteins, but some code for information that decides when and how nearby genes are turned on or off, and others code for RNA molecules that aid in the manufacture of proteins or in the movement of proteins from one part of a cell to another. Only about five percent of our DNA seems to have such an identifiable function, which is probably more a reflection of our ignorance than a lack of function for the remaining ninety-five percent.

Genotype: The collection of genes possessed by an organism. These genes interact with each other and the environment to produce the organism's phenotype, its physical appearance and behavior.

Hemoglobin: The pigment-containing protein in red blood corpuscles that carries oxygen to tissues. Hemoglobin genes come in many allelic forms, some of which cause disease.

Heritability: The degree to which a character can be selected for on the basis of its phenotype. If you select for heavier cattle by breeding only the heaviest cattle but get nowhere, then the heritability of that character is zero. But if the offspring are as heavy as the selected parents, then the heritability is a hundred percent. All real heritabilities lie somewhere in between these extremes.

Heterozygote: An organism with two different alleles of a gene, one from each parent.

Homo erectus: A large-brained hominid that first appeared in the fossil record in Africa and East Asia almost two million years ago. (The record of stone tools suggests that it may have made its debut 2.5 million years ago.) In Java there is evidence that it survived almost to the present time.

Homo habilis: The first member of the genus *Homo*, a heterogeneous collection of apparent transitional forms between Australopithecines and *Homo erectus*, with brains roughly intermediate in size. They lived about two million years ago. The fossil record is confusing, however, and much evolutionary ferment was likely going on at that time. Exactly who was related to who is still a matter of conjecture.

Homozygote: A person or other organism with two copies of one allele of a gene. A homozygote for sickle cell anemia, for example, received one copy of the mutant allele from one parent and the second copy from the other parent.

Hypoxia: The effects caused by lack of oxygen. The bodies of Tibetans, well adapted to high altitudes, can respond more readily to hypoxic conditions than those of lowlanders.

Identical twins: Twins that result when, early in development, a single fertilized zygote splits into two. They have identical genotypes, though through environmental vicissitudes they may not have identical phenotypes. (*See* fraternal twins.)

Leber's optic neuropathy: One of a growing collection of diseases caused by defects in mitochondrial genes. In this disease, metabolically highly active nerve cells in the fovea of the eye die prematurely, causing a blind spot in the center of the field of vision.

Melanin: One of several types of pigment, ranging from yellow and red to black. Melanin is also made in the substantia nigra of the brain, though by a different biochemical pathway from that made in the skin.

Microsatellite: A short repeated bit of DNA, often taking a form like . . . CAGCAGCAGCAG. . . . It can readily grow or shrink in length through mutation. Microsatellites are found on all our chromosomes. Little is yet known about their functions, but many of them may be involved in gene regulation.

Mitochondrion: A structure in a cell's cytoplasm that is responsible for much of its energy production. Mitochondria are descendants of bacteria that invaded our cells two billion years ago. They have their own chromosomes, albeit so reduced in size that they no longer carry enough information to reconstruct an entire mitochondrion; genes in the nucleus must also contribute.

Mountain sickness: A disease characterized by subcutaneous and intestinal bleeding, among other symptoms. In people who live at high altitudes, the effects of mountain sickness can be cumulative and debilitating.

Mutation: A change in an organism's DNA that can be passed on to the next generation. A mutation can take many forms, ranging from a substitution of one base for another to the gain or loss of an entire set of chromosomes.

Narcolepsy: A rare condition in which people (or dogs) suddenly and inexplicably fall asleep. Several genes have been implicated in this disease, but its exact nature has yet to be worked out.

Natural selection: A natural process in which individuals best adjusted to their environment survive and pass on their genes. Darwin was the first to realize that this process ensures that organisms best able to leave offspring are the ones that will pass their characteristics on to the next generation. We now know that natural selection depends ultimately on mutations, which provide the genetic variation that selection sorts out. Although mutations appear in all populations all the time, only a tiny minority of them will aid their carrier's survival under any particular circumstances. Genetic recombination shuffles the variation in a population, providing an important source of new genotypes on which natural selection can act.

Neandertals: Heavy-browed, big-boned people who lived throughout Europe and parts of the Middle East from perhaps a hundred thousand to 28,000 years ago. Their brains were at least as large as ours, although their skulls were of a somewhat different shape. Their ancestors, the pre-Neandertals, showed a less extreme morphology; their traces have been found in Europe dating to almost a million years ago.

Nucleus: A globular structure in the center of a living cell that contains the chromosomes. RNA carries genetic information from the DNA of the nucleus to the surrounding cytoplasm, where proteins are made.

Oldowan: A tool-making tradition, named after Olduvai Gorge in Tanzania, in which flakes are struck off stones to produce simple cutting tools. Oldowan-type tools date from as long as 2.5 million years ago, and are found widely through Africa, Europe, and Asia. The process may have been invented more than once.

PCR (polymerase chain reaction): A technique for synthesizing large quantities of a particular DNA segment. Using this technique, one copy of a particular piece of DNA can be multiplied millions of times.

Phenotype: The physical or behavioral characteristics of an organism, produced by a combination of genes and environment.

Plasmodium: A tiny single-celled animal that multiplies in red blood cells and causes malaria. *Plasmodia* are transmitted from one person to another by mosquitoes.

Pleistocene: The geological epoch that encompasses the time from 1.64 million years ago to the beginning of the Holocene, 11,000 years ago. The Pleistocene was characterized by repeated ice ages. Much but not all of our evolution took place during this epoch.

Polymorphism: The ability to take on many forms. Polymorphism may be either visible—as in the great range of hair and eye colors in the human population, or less obvious—as in the phenotypically invisible but very important ABO blood groups.

Population genetics: The study of genetic variation in populations. Population geneticists attempt, by mathematical and experimental means, to disentangle the many processes—selection, chance events, migration, and mutation—that cause allele frequencies to change over time as populations evolve.

Recombination: The shuffling, in each generation, of the genetic variability in a population into new combinations, as chromosomes recombine in the course of the production of sex cells. It is a powerful force in evolution.

Retrovirus: An RNA virus that can insert its genes into the cells of its host. The best-known retrovirus is the AIDS virus, but there are many others.

RNA (ribonucleic acid): A molecule that carries information from the DNA of a cell's nucleus to the rest of the cell. It has a structure very similar to that of DNA but it is usually single-stranded.

Selective sweep: A natural process in which one allele of a gene suddenly becomes extremely advantageous and sweeps through a population, dragging along quite unrelated alleles that simply happen to be nearby on the chromosome. Sweeps have the potential to bring about large, rapid changes in a gene pool, though their role in human evolutionary history is unclear.

Sexual selection: An important subcategory of natural selection, in which individual organisms compete to be chosen as mates. Competition for mates may be so strong, however, that it actually drives organisms in the direction of lower Darwinian fitness—elaborate mating displays and showy colors may increase the likelihood of being eaten, for example. If sexual selection gets out of hand, Darwinian selection is sure to correct the problem.

Sickle cell anemia: A genetic disease, common in Africa and the Middle East, that causes blood cells to assume a sickle shape when the oxygen level drops. The gene for the disease, when heterozygous, confers some limited protection against malaria, but it causes a severe anemia and many other problems when it is homozygous.

Size bottleneck: *See* bottleneck.

Sociobiology: A field of study that assumes that human behaviors, and those of other animals, have a genetic component and therefore an evolutionary history. More recently, this field has given rise to the subdisciplines of evolutionary psychology and Darwinian medicine.

Telomerase: A recently discovered enzyme that maintains the ends of chromosomes. Cells in tissue culture in which the level of telomerase is artificially boosted do not die as quickly as normal cells.

Thalassemia: From the Greek for "the sea"; a genetic disease that can cause severe anemia when alleles for it are homozygous. It is the result of a block in the synthesis of a component of hemoglobin. Heterozygotes have some protection against malaria. Different forms of thalassemia are common in the Mediterranean and Southeast Asia.

Transgenic organism: An organism in which a gene from another organism has been inserted. Molecular biologists now have this capability; already pigs, sheep, and cattle are being used to manufacture clinically useful human proteins. Where all this will eventually lead, of course, knows God, as *Time* magazine would have said.

Transition: A mutation change in which DNA bases of similar sizes are substituted one for the other. A and G are both bases known as purines; a change from an A to a G or vice versa is a transition. Similarly, T and C are pyrimidines; exchanges between these bases are also transitions.

Transversion: A mutation change involving a substitution of a pyrimidine for a purine or vice versa; A to T or vice versa, for example. Transversions are less likely to happen than transitions.

Tyrosinase: An enzyme responsible for making dark pigment in some of our cells.

Williams syndrome: A condition caused by a small chromosomal deletion, often resulting in mental retardation but sparing verbal ability.

INDEX